LIFE AS WE DO NOT KNOW IT

Peter Douglas Ward is professor of biology, professor of earth and space sciences, and adjunct professor of astronomy at the University of Washington in Seattle. Since earning his Ph.D. in 1976, Ward has published more than one hundred scientific papers dealing with paleontological, zoological, and astronomical topics. He is the author of twelve books, including *Gorgon, Future Evolution,* and *The End of Evolution,* and is the coauthor of *Rare Earth* and *The Life and Death of Planet Earth*. Ward has also appeared in numerous TV documentaries for PBS, the Discovery Channel, and The Learning Channel. He lives in Seattle, Washington.

Life as We Do Not Know It

The NASA search for

(and synthesis of) alien life

Peter D. Ward

PENGUIN BOOKS

For Don Brownlee, Joseph Kirschvink, Geoff Garrison, Ken Williford,

Jim Haggart, Ken MacLeod, and Roger Smith: Colleagues

And Patrick, Nicholas, and Christine,

of course

PENGUIN BOOKS
Published by the Penguin Group
Penguin Group (USA) Inc., 375 Hudson Street, New York, New York 10014, U.S.A.
Penguin Group (Canada), 90 Eglinton Avenue East, Suite 700, Toronto,
Ontario, Canada M4P 2Y3 (a division of Pearson Penguin Canada Inc.)
Penguin Books Ltd, 80 Strand, London WC2R 0RL, England
Penguin Ireland, 25 St Stephen's Green, Dublin 2, Ireland (a division of Penguin Books Ltd)
Penguin Group (Australia), 250 Camberwell Road, Camberwell,
Victoria 3124, Australia (a division of Pearson Australia Group Pty Ltd)
Penguin Books India Pvt Ltd, 11 Community Centre,
Panchsheel Park, New Delhi–110 017, India
Penguin Group (NZ), 67 Apollo Drive, Mairangi Bay,
Auckland 1311, New Zealand (a division of Pearson New Zealand Ltd)
Penguin Books (South Africa) (Pty) Ltd, 24 Sturdee Avenue,
Rosebank, Johannesburg 2196, South Africa

Penguin Books Ltd, Registered Offices: 80 Strand, London WC2R 0RL, England

First published in the United States of America by Viking Penguin,
a member of Penguin Group (USA) Inc. 2005
Published in Penguin Books 2007

10 9 8 7 6 5 4 3 2 1

Photographs from NASA

ISBN 0-670-03458-4 (hc.)
ISBN 978-0-14-303849-8
CIP data available

Printed in the United States of America
Set in Minion   Designed by Francesca Belanger

Preface

There are four goals that I am trying to accomplish in the following pages.

The first is to bring the public up to date on progress in the relatively new scientific field called astrobiology, the study of life in the universe, which of course includes the study of life on Earth. An astonishing rush of twenty-first-century discovery has rendered obsolete most books on this subject, even those published in the last five years. Some of the new knowledge that has made so many excellent titles so out-of-date has been accumulated by space probes, and some by scientists feverishly working here on Earth. The sum of these robotic and human efforts has given us a new and much more positive view of prospects for life beyond Earth in our solar system, as well as a wealth of new insight into how earth life arose and what it is. In pages to come I will show why there is such new optimism about life beyond Earth, and optimism as well about a once seemingly impossible task, the synthesis of artificial life here on Earth. Yet as wonderful as this story of new discovery is, by itself it would not have been enough to tempt me away from other research projects just to write such a book. However, without this updating I could not coherently accomplish my other, more interesting goals, all of which deal with presenting new ideas and research that I either am involved with or help fund through a large NASA grant that I administer.

One such new view informs my second goal: to redefine that elusive property that we call life and then show that mainstream biology has inadequately addressed this issue in its currently narrow view of what life is or is not. I propose that objects now considered unliving, such as viruses, as the best (but not sole) example, are indeed alive. With this leap I will then propose a new classification of life and justify why such an undertaking is necessary at this time. This finally leads to a presentation of a new, comprehensive scenario for the evolutionary pathway to, and environment of, life's origin on Earth. And it is not just a wider view of what life is that requires a new classification of "life." Astonishing progress in the synthesis of artificial life in the first few years of the new century (coupled with the realization that any discovery of alien life on another planet or moon in our solar system probably cannot be classified using our current model of life's evolution, unless it is earth life that has traveled there through space) has convinced me that the so-called tree of life, the current model of the evolution of life as we know it, must be amended in fundamental ways. Like a newly bought house with a few extra bedrooms, my new classification methodology now has room to house any newly discovered life. Thus, if we ever do find life in the solar system and *when,* not *if,* we complete the artificial synthesis of life in the laboratory of organisms that cannot be classified among our familiar earth life, there will be a way to classify it.

All the housekeeping of Goal 2 is necessary to arrive finally at my third goal: a rational look at what alien life might be like. Only by looking at what earth life is may we show what it is *not.* Goal 3 is intended to show what other chemical assemblages might arrive at the state that we call life.

My fourth and final goal is to take the newly available information about potentially habitable environments elsewhere in the solar system and couple it with the new hypothesis on the evolution and first formation of earth life, as well as with the examination of alternative forms of potentially permissible life, to make informed proposals about the kinds of life that we might find out there in space. Thus, on the basis of new discoveries from Venus, Mars, Europa, and

Titan, I will ask three questions about each of these planets or moons:

- Could any known kind of earth life live there (or have lived there in the past), and if so, what kind?
- Could any chemically permissible kind of nonearth life possibly live (or have lived) there, and how would it be classified?
- Is there a scientifically permissible way that such life could evolve or be transported there in the first place?

I am sure that to some of my fellow scientists the new hypotheses proposed in the following pages will be heresy. For example, ever since Darwin, biologists have universally agreed that there is but one kind of life on Earth, what we might call DNA life. Second, everyone except UFO loonies and science fiction moviemakers knows that we humans have never encountered aliens. I will show that both these "truths" are false. We are just beginning to realize the wondrous diversity of life that is possible in this biological universe. Most of it might be called life as we do not know it, the overall subject of this book.

Perhaps life has evolved only once. But there may be other life, perhaps it's even common. Perhaps, even, elsewhere in our solar system. If so, surely it will not all be like earth life. What will alien life be like? Like the famous story, will it be the lady or the tiger—or something so different from us that no words will do?

Acknowledgments

It is hard to know where to start. First and foremost, I must thank NASA and the NASA Astrobiology Institute, Bruce Runnegar, director, and Rose Grymes, associate director. I have used new information from many branches of NASA, and I thank its administrator, Sean O'Keefe, as well as Mike Meyers and Carl Pilcher. The other principal investigators of the institute have been unfailing in their support. Special thanks to my writing partner and friend, Don Brownlee. Joe Kirschvink shared insight and unpublished information, as did Llyd Wells, Roger Buick, Barry Blumberg, Jim Staley, Woody Sullivan. Larry Shea read the whole manuscript and made valuable comments. Major thanks to John Baross and Steve Benner, without whom the writing of this book would have not been possible. Thanks too to my editor, David Cashion, for dealing with my many mental illnesses, and to my family, for putting up with my long hours locked up writing when we could have been out waterskiing.

Contents

Introduction

His High Exaltedness, the great Jabba the Hutt, has decreed that you are to be terminated immediately.

—Some guy in a funny costume, to Harrison Ford, in *Star Wars*

It is now one of the most famous barroom scenes in movie history. A dusty town, the darkened bar filled with riffraff (all well-armed and hair-triggered, of course), the wailing loose women draped about, but this is not Deadwood, or Tombstone, or any recognizable place in the Old West. This place comes from the imagination of George Lucas. It is on some backwater planet in a galaxy far, far away, and aliens populate the bar. *Lots* of different kinds of aliens, by the disparate look of them, aliens independently spawned on a great many different planets. The message was clear. The universe is packed with alien scum, alien heroes, and aliens in between. Or so the filmmakers would have us believe. And believe we did, and still do—untold millions of us.

This iconic scene from the original *Star Wars* is as good a place as any to begin a book about life, of which earth life and alien life are the two kinds. Modern science has shown that there are certainly enough places to evolve all those aliens found in that nasty bar on the desert planet named Tatooine. Consider, there are perhaps four hundred billion stars in our Milky Way galaxy and perhaps twice that number in our sister galaxy Andromeda. Most stars in galaxies *have* planets, we are pretty sure, on the basis of the spectacular new work of such planet finders as Geoff Marcy and Paul Butler. So many planets in our galaxy and so many galaxies. Ours is but one of *billions.* The numbers make it a safe bet that planet Earth is not the only one with life or even intelligent life. So what might all these other aliens be like? Exactly what *is* an alien anyway?

Let's return to the bar in the first *Star Wars.* Obi-wan, Han, Luke, and that big Wookie are surrounded by a variety of aliens, each sporting what at first glance seems to be a radically different body plan from ours. Some have more eyes or different mouthparts, and lots of them have different hands. Yikes, the honchos from SETI (Search for Extraterrestrial Intelligence) here on modern earth would be falling all over themselves for joy. All that money that they spent was not on a fool's errand! But on further inspection, all the aliens in that bar really don't look so different from us after all. In fact, you could put a human inside every one of those rubber suits. Is this what makes an alien: a slight (or even major) difference of external body morphology? If we found any one of these creatures on some deserted jungle island or high plateau in the Brazilian rain forest, for example, would we immediately call for Tommy Lee Jones, Will Smith, and the other agents from *Men in Black*? The truth is, alien life, which we in the astrobiology community have dubbed life as we do not know it, needs more than a picture to be recognized as such. Alien life may or may not have fingers like ours, but it will surely differ in something far more important than body parts. Even blue blood would not qualify one as being an alien; our earthly cephalopods bleed blue, for instance. Something far more profoundly different must be present to qualify a life-form as alien to our familiar earth life. A true alien will differ from us in the most important and unifying characteristic of earth life: whether it has and uses DNA like ours, and even if it has DNA, whether it uses the same genetic code as we do and makes proteins with the same amino acids, among many, many other possible differences that are potentially possible.

Where will we have to go to find an alien if we use *that* kind of definition for alien life? Mars? Titan or Europa? The next star system? How about here on Earth? Surely this latter suggestion is whimsy. Where could aliens be hiding here on Earth, other than in secret Nevada military bases? *In fact, the nearest aliens might be right here, under our noses, even in our noses.*

This seems like a startling heresy. Since Darwin famously said it was so, all biology texts proclaim that *all* earth life came from a single source, that all is united by common ancestry. In other words, ever since Darwin it has been scientific gospel that all life on Earth is of the same basic type: No aliens here, thank you! However, we do not know if that is, in fact, true.

Here is the best that science can tell us: There may be lots of "aliens" (life as we do not know it) lurking here on Earth. How would we know if we ran into such life? Only the equivalent of a blood test will tell us. Here we run into the most interesting irony. We use DNA tests to establish the family tree of life on Earth. Aliens, even terrestrial aliens (an odd concept), would not show up as *life* under such tests. Maybe we need the men in black, right here on Earth, more than we know.

What we do know—with our heads, if not our hearts—is that each of those supposed aliens in the *Star Wars* bar is but a human in a rubber suit or the mirage produced by fingers running across some special effects–producing keyboard. But it is possible—not probable, but possible—that even if every one of those alien *Star Wars* desperadoes was spawned on a planet orbiting a star other than our sun, *none* of them would be an alien because it is possible *that all life in the Milky Way galaxy has DNA identical to ours.* Hence these so-called aliens are really different from earth life only in their comparative morphology, not the real stuff that makes one an alien. DNA might be one way to make life—*or the only way.* Life in fact may have spread throughout the galaxy by natural methods and is united in its origin and DNA, a hypothesis more than a century old known as panspermia. At this time, stuck as we are in this lonely and separated star system, we cannot know. Sadly, we may never know—about life on planets orbiting other nearby stars, that is. The distances may be too great, even with our amazing technology. There may never be a warp drive, or faster than light speed drive, or whatever it is called in the movies. Perhaps the stars will always be beyond our reach, something never seen in the sci-fi movies. But knowing if there is life on our own planetary companions, and indeed if there is alien life here, now, on our own planet, is another matter. We *will* know, and soon, if there are aliens in the solar system. NASA will see to that. It is busy spreading its probes across the reaches of our sun's planets even now: Mercury, Venus, Mars, Saturn, Pluto—all are the current targets, and soon we shall return to Jupiter.

So if we cannot (yet) *find* aliens, why not *make* them, while we wait for the first Mars sample return mission, scheduled to bring soil from Mars back to our earth in but a few short years from now? Here our sci-fi space epic morphs into *Frankenstein,* only updated. As I write this, several labs around the world not only are furiously trying to make life—life, in fact, as we do not know it—*but have already done so*—made alien life, in fact. So out of the bag, damn cat. There *are* aliens in existence, if we count life that does not share our specific kind of DNA, amino acids, and proteins as being an alien. That too is a subject of this book.

The time is thus right to take a new look at what an alien might be. A new and concerted effort by NASA has made respectable what was once fringe and pseudoscience. In September 2002 and again in May 2004, some of the best scientists on the planet were assembled at the behest of NASA and the National Academy of Sciences to compare notes about alien life. These scientists asked brash new questions, and not a few wondered if indeed DNA-based life is the unique type of life on Earth. They asked if there might be either fossil or extant life in our solar system that also qualifies as "life as we do not know it." They asked too if NASA and the other space agencies, which have already poured billions of dollars into current and upcoming missions (to Mercury, Venus, Mars, and the Saturnian system), are building the right machines, visiting the right planets and moons. They are posing brazen questions about the possibilities of alien life. There is reason for optimism if you, like me, hanker to *really* know if we are alone. On the basis of masses of new data and insights, the solar system seems a far more biologically friendly place than was assumed even in the last years of the twentieth century.

Thus the time seems propitious. But why did I choose to write this narrative, and what can I bring to the subject? It takes only a quick look at the bookshelf in my office to find a score of reasons for why *not*. In the past decades there have been numerous and monumental books on life's origin and definition, by a parade of humanity's most gifted: Erwin Schrödinger's *What Is Life?* Jacques Monad's *Chance and Necessity*; Francis Crick's *Life Itself: Its Origin and Nature*; Christian de Duve's *Vital Dust*; Paul Davies's incomparable *The Fifth Miracle*; and Freeman Dyson's *Origins of Life,* among many, many others. I make no pretense to be in the intellectual league of these luminaries. But all these books are about how life formed on Earth (and about what life is). And while there is also a huge group of books about the environments on the planets and moons of our solar system where life might begin, there are few books that are a fusion of the two, using the knowledge of what earth life is and how it formed on Earth to inform us about the probabilities of life in the solar system. My sense is that progress can be made at this moment by a new analysis that is informed by both topics.

I also come to the debate as an acknowledged skeptic about alien life, and who better than a doubting Thomas to take a dispassionate look at life in our solar system? Near the end of the last century I spent three years writing a book with my colleague Don Brownlee about the potential frequency of aliens. *Rare Earth: Why Complex Life Is Uncommon in the Universe* was published in the first days of this new century. The center of that book is what we call the Rare Earth Hypothesis, which is really two hypotheses. First, microbial life is relatively easy to evolve from nonlife; has a wide range of tolerance in such conditions as temperature, pH, and pressure; and therefore should be found widely throughout habitable planets and moons in the cosmos. Second, most such planets and moons remain habitable for only short times or are never suitable for higher life to live on, and thus complex life (which invariably seems to be more fragile than simple life and requires ever more and narrow environmental conditions to survive), the equivalent of our animals and higher plants, will be very rare indeed.

In the blissful, now sadly departed pre-9/11, pre–Iraq War, and pre-SARS days, our book, for whatever mysterious reason, became widely publicized, aided in no small way by the howls from the SETI organization, which, like the science fiction industry, depends on a belief in aliens for its economic viability and thus saw our book as a personal affront and attack. From that point onward I was (and continue to be) viewed as the number one skeptic about life in space, and I am usually wildly misrepresented as suggesting that life, or even just intelligent life, is unique to the earth. That is ironic, and it comes from those unfamiliar with the book, for *Rare Earth* was more about why simple life should be common than about life's rarity. Nevertheless, I do approach the subject of life in space with conservatism.

Why is it that there seems little hope that we will find Wookies (or their animal-like ilk) on every planet—Mars, for instance? Could H. G. Wells, SETI, and a thousand followers positing inimical Martians and untold other nasty (and biologically complex) aliens have gotten it *so* wrong? I do think so—because they have looked up at the stars, rather than at the rocks beneath their feet for primacy of information about life in space. The reasoning behind the Rare Earth Hypothesis came not so much from astronomy (although there was plenty of it necessarily there) as from new information from oceanography, geology, and paleontology—important players in the then newly forming science called astrobiology. In *Rare Earth,* Brownlee and I suggested that while the sheer number of stars with their inevitable planets would make the existence of at least some other intelligent civilizations a near inevitability, their probable small numbers would space them apart in the cosmos at such vast distances that it would be unlikely that one intelligent civilization would ever even detect the *presence* of another, let alone meet it face-to-face, or face–to–whatever an alien meets with. We defended this view by describing the geological and planetary mechanisms that allow the earth to maintain its life, such as the overriding importance of what can be thought of as a planetary thermostat of planet Earth, a geological mechanism involving plate tectonics and the movement of carbon from rocks to atmosphere and back again. We described how the combination of stable temperature for billions of years and a low number of mass extinctions (which can be thought of as potential planetary sterilization events) in the past had combined to allow the earth the time necessary for the evolution of animals and higher plants from the original microbial stocks. We asked how similar conditions might persist on other worlds, and we pondered the set of conditions that might allow other worlds to maintain long-term temperature and atmospheric stability. It seems that many separate conditions are necessary for a planet to remain habitable for long periods of time.

But critics of the book (and they remain legion!) rightly pointed out a most crucial point, a variation on what is known as the anthropic principle. *Rare Earth* dealt *only* with a narrow segment of life, something that could be called earth life, sometimes referred to as life as we know it or DNA life. This type of life, the argument goes, would necessarily require a planet much like Earth for its survival, and all acknowledge that there are probably few exact duplicates of Dear Old Earth. Yet the universe is large, and our solar system and the earth are small. The very immensity of the universe, with its myriad galaxies, *must* allow for an enormous diversity of life, according to those who believe that there are a vast number of aliens and alien civilizations in space. Not *diversity of life,* as we understand that phrase applied to life on this planet. All biologists agree that there are millions of species on Earth but that all sprang from a single common ancestor and thus are but one type of life: DNA life. Yet it is surely possible that there must be a multitude of *types* of life differing in their most fundamental properties of chemistry and metabolism. Such life would not have DNA, or perhaps even carbon as a constituent, like the oft-discussed staple of science fiction silicon life. Such life indeed might be life as we do not know it.

With the vastness of the universe and the large number of distinct chemical elements found throughout it, surely it is reasonable to suppose that all manners of life are possible. But is this really true? How diverse can life really be? How about boron life? Or tin life? Or life without any matter at all, like some form of pure energy life? With a bit of imagination almost any type of life can be suggested. But life is not a simple proposition. It may be the most complex of all chemical systems in the universe. It seems to me that there is a reasonable possibility that there might *not* be many ways for life to be constituted at a fundamental chemical level. In fact there might not be *anything* except carbon-based life-forms. And in an even more restricted sense, DNA might not be just one way *but the only way*. So, if not DNA, what? Perhaps all life uses DNA, perhaps with a different "language" or syntax in its meaning, but DNA nevertheless. This point of view cannot be disproved at the present time, a sad commentary on how little work about the chemistry of life has actually been accomplished.

To ask about the diversity of types of life requires that one looks at another, even more difficult question. How could anything as complex as DNA have come about naturally, on Earth or any planet? It is not enough simply to arrive at a chemically permissible form of life. Any discussion of aliens must ask how they originated.

How life formed on Earth is indeed one of the great (some say the greatest) scientific mysteries, a problem perhaps even more important than the mystery of how common life is in the universe (indeed, the latter question is an offshoot of the first). There are currently two schools of thought concerning this origin of life problem. First, some deity did it. That one cannot be disproved with science, so has no further place in our story here. Second, life formed on Earth (or was transported here, but this just begs the question of where and how it first formed) through some chemical pathway within a specific environment at a specific time. This view has much scientific information to support it. Many scientists now believe that life might be an inevitable consequence of this particular universe that we live in. Life appeared on Earth relatively early in earth history, certainly no later (and perhaps much, much earlier) than about 3.7 billion years ago on a planet that coalesced into existence about 4.6 billion years ago.

This suggests either that our planet was very lucky or that life—at least as it was on a planet like Earth early in its history—is not so hard to make.

These two threads—what life could be and how it could form—are thus the intertwined themes that I shall explore in the pages to come.

Finally, I am in an interesting position to tell this story. In 2001, on our second try, the University of Washington won a highly competitive five-year grant from NASA to join the NASA Astrobiology Institute. I was (and am) the principal investigator of the University of Washington node of what we call NAI (made up of fifteen research institutions around the world), and in this position—being a member of the NASA organization, yet not being under the umbrella of the organization to the point that I cannot speak about both the warts and the wonders of NASA efforts—I have seen from a wonderful vantage many of the discoveries and concepts that will make up this book. A not insignificant part of the work to be discussed has been funded by the grant that I administer, and some of it is my own. Many of the people who have made contributions important to my narrative are members of the University of Washington NAI team, and if it becomes tiresome to read about yet another discovery from that team, then they have done their job well.

It has been wondrous to see the workings of creation (as the laws of physics and chemistry might be called) and know that there must have been multiple, and separate, creations, all the workings of planetary chemistry sets—some of them, perhaps, the best thing ever evolved by our species. Or the most dangerous, if artificial life eventually comes back to bite the hand that invented it.

One of the most important messages that I wish to present in this book is that there has been significant progress in the artificial synthesis of life. While pundits routinely pontificate that science will never succeed in creating life from scratch in some equivalent of a test tube, quiet progress on many fronts begs to differ. The question is thus not, *Can* we make life, but *should* we? Should we make life that has never evolved through natural selection, life using chemistry that is a nose thumb to Darwin? We are in the post-9/11 world, of course, so we have to ask if this new life can be dangerous: Can it be made into a weapon? Can it be made into a tool that does the *opposite* of weaponry: helps humanity with food, or with pollution control, or in fighting disease? Or should it be done simply to show that life other than that evolved on Earth is indeed possible? What if we could build a life-form that could live at −150°C—essentially the temperature found on the moons of Jupiter and Saturn—and thus demonstrate that life could indeed live in these cold freezers of the solar system? This would prove that life could live in such cold places, and we would be ready to look for it. These would *really* be aliens. This is something that I wake up worrying about, in the dark of the early morning, the time of nightmares, for I am working toward the construction of such aliens. What if it mutates, takes over the Arctic, and then spreads south toward the winters of our populated world? Should scientists play God, even if it is a Jack Frost god?

Could there be earth life on any or all of these bodies beyond Earth? And if not earth life, any—or what—kind of life? In a wonderful story by Robert Sheckley, a poor human astronaut stranded on Mars stumbles onto an abandoned (and somewhat intelligent) Martian village. By shaking out some loose crumbs from his pockets, he makes the village understand what sort of food (and water) he needs to stay alive. But this type of material is in very short supply on Mars, and in trying to make the food and water that he needs, the village starts to destroy itself. The punch line of the story is that the earthling gives himself up to the village and is transformed into a Martian, who no longer needs water. There is a message here. We too often look for earthlike conditions, because that is what is needed for earth life. What about the burning worlds like Mercury and Venus? Or the frigid wastes on Mars, where there is virtually no liquid water, temperatures are forever below zero, and even the soil contains antibiotic chemicals? What would be the ideal chemistry of life? Or on Titan, where it is even colder and harsher? What sort of life, if any, would thrive there? What about farther in space, on a comet, perhaps? Each of these places has been the focus of a recent NASA mission of some sort, with only Europa not visited in the last year. For each, there is startling new information that pertains to the question of life in the universe. Earth life works wonderfully well on Earth. But elsewhere?

We may never find Wookies, but what we find, and make, will ultimately be far more interesting than a tall man in a hairy suit. So let us begin.

Chapter 1

What Is Life?

From a commonsense viewpoint, nothing seems easier than to tell what is alive and what is not.

—Gerald Feinberg and Robert Shapiro, *Life Beyond Earth*

The small submarine headed down toward the blackness of the deep sea bottom, thousands of feet below, and the cramped men inside could only wait out the seemingly endless descent, passing the time by peering through the *Alvin*'s thick glass portholes. The voyage began in the sunlit portions of a sea filled with life modern in aspect but then descended back through the ages, for the deep sea is the home of many living fossils, species of great antiquity. Finally the voyage landed at the site where life may first have started, a place that might still harbor microbes that were present when DNA life first arose. Such a place might harbor other life as well, the predecessors of our familiar DNA life. But at what point as we go back do we reach the transition from life to nonlife? That particular question was about to become even more problematical on this day in 1978.

At first, the rich fauna of the daylit upper reaches of the subtropical Pacific surrounded the submarine: clouds of plankton, most evolved from the Cretaceous period on; schooling small fish and the larger piscine predators that pursued them, relics too of the

Cretaceous period; umbrellas of jellyfish and their ilk, from vastly older stocks, but made up of species that might be very recently evolved; the shooting forms of arrowworms; the darting and resting of crustaceans that seem even more modern than the fish in these sunlit seas. Farther downward the submarine descended, and the brightness and color of the sea changed, a gradual journey through the spectrum of blue in all its shades toward ever-deeper hues. Now life was less noticeable and different in aspect. The sardine shape of the fish changed, as did the look of the invertebrates, and now, at least to a paleontologist, the world looked more Mesozoic than otherwise. An occasional squid appeared, and even these no longer looked like their familiar surface-dwelling cousins. Long tentacles draped from the head regions in some, while others had squat, ammonia-filled bodies resembling tiny hot-air balloons. These exotic squid were probably the closest living relatives in both ancestry and ecology to the vast race of the extinct ammonites, poor victims of the end-Cretaceous asteroid, and they themselves were of great age. Here in the mid-water now, in a place as far from the bottom as from the top of the sea, there lived a fauna that depended on flotation sacs to keep it permanently suspended—a fauna from antiquity. Through the windows of the small sub, deeper now, more than a mile beneath the surface, the scientists could see undulating floaters, or slowly swimming invertebrates giving off rainbows of shimmering color. Deeper still, approaching two miles deep, small lights, like drifting stars, amid the surrounding pastures of life, not unlike fireflies at dusk on a warm Ohio evening, began to appear in the darkness. The larger carnivores—the peculiar fish and squid of these great depths—some amazingly grotesque with huge mouths lined with sharp teeth, on occasion passed by the windows. All were countershaded, with light-lit bellies and dark upper surfaces, so that anything above them would see only black, and anything below would not notice their silhouettes against the faint light of the far-distant surface. All had a shape that comes to us from the distant Paleozoic, when the platy placoderms and armored arthrodires known from spectacular fossils recovered in the Cleveland shale or Old Red Sandstone evolved the peculiar heterocercal or reversed shark-shaped tails that are found only on these deep-living species and on the fossils from the four-hundred-million-year-old seas of ancient planet Earth.

As the deep blue of the tropical sea lost finally all color and was replaced by velvet night, the great searchlights of the *Alvin* switched on, bringing bright light to a region that had been unlit for millions of years, and the sub finally reached the bottom, a journey down, and one seemingly back in time, to the dawn of life on Earth. An hour had passed since the submergence. After the seemingly endless descent, when the men's imaginations had already cataloged the many ways that the submarine could fail its fragile inhabitants in the pressurized and cold depths, the bottom was finally spotted. Stark, lifeless, it resembled all the sea bottoms from the time before the Cambrian explosion, that moment 550 million years ago when the animal phyla sprang into existence. For 3.5 billion years prior to that moment all sea bottoms had been barren of any life but microbes, barren like this one.

For a time the *Alvin* glided over this desert sea bottom, a vision, perhaps, of what the deep-sea bottom of Europa might look like, surely also a smooth expanse of sediment in absolute darkness, but is it, or even this sea bottom, really lifeless? Then, shockingly, the smooth mud of this deep Pacific Ocean bottom gave way to a lithic landscape. A tangled rocky wilderness, the land of the submarine volcanic vents, lay illuminated below the *Alvin*. The bottom was like a disordered junkyard, with vast fields of pillow lava and twisted toothpaste squeezes of now-solid rock covered with a patina of sediment. The *Alvin* was about fifteen feet above the bottom now as it powered over the endless fields of volcanic rock, when its startled humans found themselves among a sudden profusion of animal life, one unlike that of the surface regions. It was a vent fauna, the strange animal fauna that had first been seen only two years before during dives in the Galápagos Islands. But those dives had been on a bottom far more peaceful than this one. This water became hazier, filled with dustlike particles and larger flocculation of repulsive-looking slime. Now there were tube worms, and white clams and crabs, and catalogs of other invertebrates unknown to humanity. In the worms and crabs and clams there was no doubt that the *Alvin* had found life—weird life, of course, but unambiguous vessels of earth life. But what of the floating slime, the white snowflakes that clouded what had been a pristine sea only minutes before? What was this material? Was this life?

The rocky bottom became more dissected, with walls and deep, narrow canyons appearing. The rocks in this new canyon land were covered with brown growth—mats of microbes?—and while the occupants of the small sub could not know it, much of the microbial slime that they viewed amid this deep rocky rubble was, like the larger animals, composed of living fossils, but in this case, species not of some Paleozoic age but of a far more ancient time, the time of the earth's youth, time measured in billions, not millions, of years. Suddenly and unexpectedly a tall spire of rock appeared dead ahead. It was covered with life, but the scientists looked at something far more arresting: This vertical rock column was belching shimmering black smoke into the dark water at a prodigious rate. Beyond, in the murky water, other tall chimneys could be dimly seen. Some were three or four stories tall. There was animal life here, most notably large white clams and spectacular tube worms, forms that had been seen on the *Alvin*'s dives off the Galápagos Islands, but it was not the animals that so startled the astonished scientists in the small submarine. The submarine had seemingly entered a Tim Burton nightmare, the Gotham version of Industrial Revolution England, in which tall black chimneys of rock spewed blacker smoke into the clear seawater over a dingy dark town of jumbled rocky tenements. The scientists had reached the industrial heart of the planet. Humans, for the first time, were seeing a landscape of the black smokers, and a whole new vision of how life first formed on Earth, and perhaps anywhere else, was conceived.

The black smokers had to be more closely investigated. The cramped scientists aboard the *Alvin* had the pilot use the vessel's mechanical arm to knock off the top of one of the rocky smokestacks, and in response the decapitated chimney belched its noxious black liquid at an even faster rate. Temperature probes were cautiously inserted near the beheaded smoker. The black smoke (actually a superheated fluid coming from deep within the earth rather than real smoke) was found to be hot—nearly 100°F in temperature, in fact. This was far hotter than the hottest temperature recorded in the first of the deep-sea vents found in the Galápagos Islands two years prior to this dive off Mexico, which had been about 60°F. The scientists suspected that the temperature gauge was somehow giving false readings. But repeated analyses showed that the hot temperatures were real. How hot was the fluid before it left these tall lithic chimneys?

Later, when the *Alvin* was back on its mother ship and the maintenance crew could work on its instrument package, the temperature probe was examined. Its plastic end was melted. The excited scientists did a quick experiment to find the melting temperature of the tough plastic. It was found to be over 350°F, well in excess of the 212°F temperature at which water boils. Superheated water was apparently emanating from these strange geological chimneys, the black smokers. Better thermometers were quickly built onto the *Alvin* probe system. On subsequent dives the researchers could only read their temperature data with heads shaking in astonishment. The temperature of the black fluid coming from within the smokers was measured at over 600°F. The scientists were stunned. No one had foreseen this. The liquid remained liquid because of the high pressure of the great depth.

Pieces of the chimneys were brought to the surface and found full of minerals permeated with sulfides. Such minerals were also known to serve as the energy sources of a large group of bacteria, thus explaining why every surface in the smokers' environment was coated with microbes. On successive dives another phenomenon was witnessed. This time it was not fire and brimstone but the proof of life. Microbial life was found to be living amid this hellish embrace, not only on but also in the smokers. A new kind of ecology was being witnessed, composed of species that could not only live in extreme environments but also actually thrive there in great numbers. Unlike all other ecosystems on planet Earth, which are ultimately powered in one way or another by sunlight mediated by photosynthesis, this assemblage of large animals and microbes had, as its ultimate energy source, nothing less than the heat of the earth, and the actual food-stuff of the strange animals was bacteria fueled not by sunlight but by hot, chemical-rich fluids coming up from Hades itself.

The environments around the deep-ocean volcanic rifts that were the home to these newly discovered ecosystems can be described by a single word, *extreme*. Extreme heat surrounded by extreme cold, all in extreme pressure and absolute darkness, and amid toxic-waste chemicals—in short, in conditions seemingly inhospitable to any living thing. Yet within these scalding cauldrons of superheated water, a rich diversity of microbial entities grew and thrived at temperatures far too hot for any animal. It made one think and reconsider the so-called rules of life and the assumptions about where life could and could not live.

Earth, apparently, has at least *two* quite different regimes of life: the traditional one known for centuries, in which plants trap sunlight, and convert CO_2 to living matter, which is then eaten by herbivores, which in turn are consumed by larger and larger carnivores, with all being infested by parasites and ultimately consumed by scavengers, and this new, second system, in which bacteria trap energy from the volcanic vents—no sunlight needed, thanks—with the bacteria taking the place of the green plants in the first and more common system. The microbes that were found to be the lowest link of the food chain in the vents lived off sulfurous compounds found within the vents. In this system the microbes were the equivalents of the plants found everywhere else on the planet's surface. Only slowly did it dawn on most marine biologists that to understand this new system of life, one first had to understand the metabolism of the microbes.

The dive described here was made in the late 1970s, and it is a good place to begin a book about life. The time of these first discoveries is really when the science of astrobiology began in earnest. The entire new suite of creatures found on these dives discredited much

Black smokers as seen from the Alvin *on the deep Pacific Ocean sea bottom.* (Courtesy NOAA/ALVIN)

"common knowledge" about earth life and confounded the then complacent field of biology, which had grown smug in its belief that its practitioners had pretty much found all the basic kinds of life there were to be found. Little did they know that within a few short years of these dives, the entire concept of life's history and classification would be overthrown by a revolution brought about by comparing genes of various creatures. But these reasons, as important as they are, are not why I decided to begin this book's story here, on this deep bottom of perpetual darkness, where temperatures are either too hot or nearly too cold for our more familiar kinds of earth life. Sometime in this century a smaller version of the *Alvin* will dive into one and perhaps two different oceans of temperature and chemistry not dissimilar to our own. Beneath the ice layers of Jupiter's moon Europa and again beneath the ice of Saturn's moon Titan, our species will, like those on the *Alvin* dives of the 1970s, send machines into an alien environment. Will there be life? This is the obvious question. Yet there is another question that is not so obvious but, as we shall see in this chapter, of equal importance, though one that might be very difficult to answer. The *Alvin,* on the first dive onto the black smokers of the deep-sea bottom, encountered a suspended cloud of floating snot that its trained biologists could not even *identify* as life until they were able to conduct high-powered examinations with state-of-the-art instruments. And that experience happened here on Earth. Will we recognize as life *any* aliens that we might encounter in the oceans of Europa and Titan? Life on Earth, even what we call the simplest life, is complicated. But have we set the bar unrealistically high in our definitions of what life might be? Could it be a slow-growing crystal or bits of clay with some scattered carbon molecules attached or even strange compounds of silicon bonded to carbon floating slowly in a pool of supercold oil on the surface of a Saturnian moon? Here we shall explore the first and most critical question facing those on planet Earth who would have the temerity to call themselves astrobiologists: What is life?

So what is life?

Some of the best minds our species has produced have wrestled with the problem without consensus. Our judgment about what is alive on some alien world will all too probably be colored by our membership in the guild of earth life, of life as we know it. But there is surely a great probability that we will eventually encounter what I shall call alien life and not even recognize it as such, or that such life exists, still undiscovered, on our Earth. While a great fear of humanity is that we will—or won't!—find aliens anywhere on Earth or in our solar system, graying heads worry more that we will stumble across life but not even recognize it as such. As we will see, we have a great deal of difficulty deciding if any number of organic-like forms on Earth are alive. So what does this say about our ability to define what is alive or not on alien worlds?

The following statement illustrates just how knotty the definition of life can be: All life-forms are composed of molecules that are not themselves alive. At what level of organization does life "kick in"? In what ways do living and nonliving matters differ?

There is a long list of really smart people who have tried to comprehend and define the nature of life (life on Earth, that is, for few of these savants were considering the bigger picture of life in the cosmos rather than simply life on Earth). The question, What is life? is even the eponymous title of several books, the most famous by the early-twentieth-century physicist Erwin Schrödinger, which provides a great starting point for this discussion. Schrödinger's short book was a landmark, not just for what was written but for who did the writing. Heretofore the fields of biology, chemistry, and physics existed in their own domains, and while the latter two had large and necessary overlaps, the world of the living had been of little concern to these more physical sciences. Schrödinger was among the first of his discipline to break down these walls. He began to think of organisms in terms of physics, and early in his book he took overt notice of the difference between the living and nonliving. While much of the book deals with the nature of heredity and mutation (the book was written twenty years before the discovery of DNA, when the nature of inheritance was still a perplexing mystery), it is late in the book when Schrödinger considers the physics of "living," writing, "Living matter evades the decay to equilibrium," and life "feeds on negative entropy." *Entropy* is the term used to describe how natural systems move from order to disorder. Schrödinger thus saw life as doing the opposite, somehow changing disorder to order, or reversing the natural trend of entropy. Hence his use of the term *negative entropy*. (How like a physicist to complicate otherwise simple things.) Life does this through metabolism, by eating, drinking, breathing, or exchanging material. Is this the key to life? Perhaps—at least to a biologist. But Schrödinger, the physicist, saw something much more profound: "That the exchange of material should be the essential thing is absurd. Any atom of nitrogen, oxygen, sulfur, etc., is as good as any other of its kind; what could be gained in exchanging them?" What, then, is that precious something that we call life, contained in our food, which keeps us from death? To Schrödinger, that is easily answered: "Every process, event, happening that is going on in nature means an increase of the entropy of the part of the world where it is going on. Thus a living organism continually increases its entropy. . . ." Life was thus the "device" by which organisms maintained themselves at fairly high levels of order by continually sucking this "orderliness" from their environment. For all his insight, some of Schrödinger's (and physics's) views about life were naive. From the physics point of view, for instance, life could be understood as a series of machines, all packed together and somehow integrated, functioning in such a way that they, and life itself, could be understood using physical laws. For a half century, then, the question, What is life? could be simply answered: Life is simply an agglomeration of machines that change disorder to order. But in the latter part of the twentieth century, biologists, chemists, and other physicists began first to question and then to amend these views.

The renaissance in understanding the scientific nature of life, stressing that there is surely more to life than biological machines and entropy, was, ironically, led by two other physicists, Paul Davies and Freeman Dyson. Davies, in his 1998 book *The Fifth Miracle*, furthers our understanding of what life is by asking, What does life *do*? If all his answers could be understood as the change in entropy as the result of biological "machines," he would have proved Schrödinger's point. But as Davies showed, there is indeed more. Here are his answers to what life "does":

Life metabolizes.

All organisms process chemicals and in so doing bring energy into their bodies. But of what use is this energy? The processing and liberation of energy by an organism are what we call metabolism, and it is the way that life harvests the negative entropy described by Schrödinger that is necessary to maintain internal order.

Life has complexity and organization.

There is no really simple life composed of but a handful (or even a few million) atoms. All life is composed of a great number of atoms arranged in intricate ways. But complexity is not enough; it is organization of this complexity that is a hallmark of life.

Life reproduces.

This one is obvious, and one could argue that a series of machines could be programmed to reproduce, but Davies makes the point not only that life must make a copy of itself but that it must make a copy of the mechanism that allows further copying; as Davies puts it, life must include a copy of the replication apparatus too.

Life develops.

Once a copy is made, life continues to change; this can be called development. Again, it is a process mediated by the machines of life but also involves processes that are unmachine-like. It is in this area that this new view of life diverges radically from the Schrödinger view.

Life evolves.

This is one of the most fundamental properties of life and one that is integral to its existence. Davies describes this characteristic as the paradox of permanence and change. Genes must replicate, and if they cannot do so with great regularity, the organism will die. On the other hand, if the replication is perfect, there will be no variability, no way that evolution through natural selection can take place. Evolution is the key to adaptation, and without adaptation there can be no life.

Life is autonomous.

This one might be the toughest to define, yet is central to being alive. An organism is autonomous, or has self-determination. But how "autonomy" is derived from the many parts and workings of an organism is still a mystery, according to Davies. Still, it is that autonomy that again separates life from machine.

It was not only the late-twentieth-century physicists who weighed in, but biologists and even astronomers as well. The great Carl Sagan famously wrestled with the question of what life is, and unlike most others thinking about this topic, who were dealing only with life as it is found on Earth, he came at the problem with a specific goal: He was interested in life beyond Earth, the life as we do not know it. In the mid-1970s Sagan had some very pragmatic reasons to better define life, for as we shall see in chapter 10, he was one of the lead scientists on an ambitious project to land large probes on Mars that NASA accomplished in 1976. Sagan's definition of life, which is still largely taken up by NASA to this day, sees life as *a chemical system capable of Darwinian evolution.*

There are three key concepts to this definition. First, we are dealing with *chemicals,* not just energy. All the *Star Trek* energy beings (how about that Q of *Star Trek*?) are thus cut out. Second, not just chemicals but also chemical *systems* are involved; thus there is an interaction among the chemicals. Finally, our chemical system must undergo Darwinian evolution, meaning that there are more individuals present in the environment than there is energy available, so some will die. Those who survive do so because they carry advantageous heritable traits that they pass on to their descendants, lending the offspring greater ability to survive.

The Sagan/NASA definition has the advantage of not confusing life with being alive. But there are problems under this definition too. For example, one gender of a species composed of two sexes cannot undergo Darwinian evolution and is therefore not alive. But there is a final note of interest about the NASA definition. If we accept it, it means that scientists we meet in the pages to come (specifically, Harvard biologist Jack Szostak and his many confederates) *have already created life in a test tube* because they have succeeded in making short RNA molecules, a chemical system, that undergo natural selection.

So when does life "begin"? By summoning all the various visions of what life is, we can say that life begins in some environment, in which chemistry—to be specific, geochemistry (the sum of the chemical reactions taking place in rocks and the air and water or other fluids above and within)—becomes intertwined with new, self-sustaining, replicating, and evolving chemical reactions among organic molecules on some planet, moon, or other heavenly body. Two different trains of chemistry join the same track, and we have life.

Defining life based on simplicity

So far we have approached the what is life question as a series of definitions. Let us come at this thorny problem from a new angle: What is the simplest assemblage of atoms that is alive? The question lets us look at the various components of a simple life-form and, system by system, ask if it is necessary to keep the whole alive. For life on Earth, we might ask: What is the simplest life-form on Earth, and what does it need to stay alive? We can take these results and ask: Is it possible to conceive of (or construct, but this will be the subject of a later chapter) an alien life-form even more primitive than the simplest life-form on Earth?

The most primitive life found on earth that we all can agree is indeed alive is a large, diverse group of microbes called bacteria. The well-known bacterium *E. coli* can serve as an exemplar of this fascinating zoo of creatures, but any number of other bacteria will do as well, and indeed as bacteria go, *E. coli* is fairly complex. Our first impression is how small this life-form is, and how simple. A rodlike shape is all there is. The outermost part of the bacterium is a cell wall with an interior plasma membrane, which encloses one large, nonpartitioned interior space. This membrane is fundamental for the life of the cell, as it separates the workings of the cell from the outside world and allows there to be interior environments that can have a different chemistry and composition from the exterior world. (Cell walls maintain cell shape and prevent bursting caused by osmotic pressure. They are porous and don't control flow in and out of the cell. The plasma membrane is semipermeable and, although delicate, provides control of substances in and out of the cell.) If we remove this wall and membrane, will the remaining chemicals still live? Clearly not. We know of no life that does not have this property of separating the workings that compose life from the larger environment. Life, then, seems to need some analogue to the cell wall that we see in all known life on Earth.

The major building block of the plasma membrane in our familiar earth life is a substance called a lipid, which has the property of being nondissolvable in liquid water. This is necessary indeed if our cell is to maintain any sort of integrity in a world such as ours where water is pervasive. The molecules making up the membrane are composed mainly of phosphorus, carbon, hydrogen, and oxygen. The fashion in which this structure is constructed gives its property of being insoluble in water. The molecules are distinguished from hydrocarbons (long chains of carbon atoms bonded with lots of hydrogen atoms) by having more oxygen. By itself, a lipid layer would inhibit water and water-soluble substances from passing through it. This would be a disaster for any life having just a lipid wall, and to deal with this, the structure making up the cell wall is more complicated. It is elongate and has one end (with a phosphate molecule attached) that can react with water, while the other end will not. This is necessary for life, for the intake of material for metabolism, and for the egress of potentially poisonous by-products of that metabolism. So most biologists who worry about the What is life? problem are pretty sure that life, any sort of life that we can define, must have some sort of membrane system equivalent to this. As we will see, even this view might be causing us to ignore more interesting assemblages of atoms that might indeed be alive, but we will have to come back to this interesting possibility later. For the moment let us humor the current dogma and ask: Is there any element other than carbon that could be used to make a membrane like that used by earth life? For the conditions that we find on Earth, the answer seems to be no.

Let us continue looking at this particular example of earth life and as miniature adventurers continue our tour of this brave earthling by traveling inward into our cell, passing through the outer membrane system that separates our bacterium from its surrounding environment. The interior of the bacterium is packed with molecules, arranged in rods, balls, sheets, and some far more complex topologies, all floating in a salty gel. If we could see these actual molecules, we would be astounded at the variety. There are about a thousand nucleic acids (these are described in much greater detail later) and over three thousand different proteins. All these are going about some sort of chemistry that, combined, makes up the process that we call life, and one of the major questions facing biology is how so many chemical processes can go on simultaneously in this one-room house.

So far so good. But let us be obnoxious astrobiologists for a moment and turn this humdrum humans-get-miniaturized plot from too many bad B movies into something more interesting. Let us arm our intrepid explorers with the same funky laser weapons (and skintight suits!) given to Raquel Welch and Stephen Boyd in the sixties howl *Fantastic Voyage,* and start shooting up the interior of our bacterium. See that floating globule over there? Zap! How about the complexly folded protein over there? Zappp! How many of the nucleic acids and proteins could we blast into goo and still leave the poor invaded cell alive? (Turning the tables on a life-form that does the same to us on occasion does produce a sense of vicarious pleasure, I must admit.) This is an experiment being conducted—using different techniques from those used by a miniature Raquel Welch, admittedly—in numerous labs in the search to find out what a minimal organism really needs to "live." It turns out that almost any microbe that we enter in this way is fabulously complicated, with a lot more machinery than is needed to be alive. If alive is a one-star hotel room in France, then these are not four-star (not that I have ever had that pleasure, sadly); these are hundred-star deluxe Ritz-Carltons. And these are not even the really deluxe models, the multicellular examples such as ourselves. Earth life: first class life all the way! Or at least the earth life that we have discovered to date. There may actually be ghettos and tenements of life that we have not yet discovered, but they are for another chapter.

Two major and different shapes confront us as we continue to travel through the interior of this tricked-out, chrome-plated earth life bacterial cell. First, there are about ten thousand individual spheres, known as ribosomes, which are distributed rather evenly throughout the cell. If we were to anthropomorphize them, ribosomes would be among the most important characters in a drama that we could title *How Earth Life Evolved*, a play that opened on the earth stage almost four billion years ago to rave notices, and wouldn't we like to know if it made it off Broadway as well, to, say, those summer theater towns of Mars, Venus, Europa, and Titan? The ribosomes are complex characters, as is demanded of any meaty role (pardon the pun here, for they do make the meat that we are composed of). They are made up of three distinct types of the nucleic acid known as RNA and about fifty kinds of protein. But they are not the stars of the show. That distinction is held by the real prima donnas, the second major morphology present: the chromosomes, long chains of DNA complexed with specific proteins.

It is the chromosome that we reach last on our journey through the cell. A long strand of DNA, the key to life on Earth, is made up of only five elements: carbon, phosphorus, nitrogen, hydrogen, and oxygen. It is this molecule that unites all *known* life on Earth and proclaims a common history and ancestry. It is this strand of DNA that governs all, ensuring that the enclosed cell is not just a bowl of chemical soup but also a functioning unit of life. On it is coded all the information necessary to keep this complex chemical factory working and alive.

So what in this cell *is* alive? Rephrased, what is life anyway?

Again, if we really do take this junk out, bit by bit, at what point does the poor victim die? Our bacterium is composed of inanimate molecules. A DNA molecule is certainly not alive in any sense that any rational person would accept. The cell itself is composed of myriad chemical workings, each, taken alone, but an inanimate reaction of chemistry. The apparent increase in order found within our cell—the loss of entropy, the negative entropy of Schrödinger's of so many years ago—does not contradict the second law of thermodynamics (which states that every chemical reaction results in a loss of energy) because it is accompanied by an increase in *disorder* around the cell itself.

Can it be said that *nothing* is alive but the whole of the cell itself? This appears to be the case for this bacterium (except if it has a few viruses within it, of course, but we will come back to this very controversial point a few paragraphs hence), and if we are to understand how life first arose, we need to find the minimum cell that can accomplish this. How small, with how few molecules and reactions and information, can it be and still have this elusive property of life? Moreover, if this is life, how can there be a similar process to living without the various molecules and chemical reactions that are part of this system as it extracts order from its surroundings, reproduces, evolves? Either we change the definition of what life is, or we are confronted by some minimum yet enormously complex system of matter and energy flow. So to make an alien, we need to change only the structural components, not life itself, it seems. Let me make the first of many obtuse conclusions: There are alien body forms, not alien life. There are many ways to cause inanimate matter to take on the property of life. But there is only one "species" of life itself.

One of the pressing problems in looking at this simple cell is that, when examined in detail, it is in no way simple. Freeman Dyson has explicitly looked at this aspect of modern life, asking: "Why is life (at least life today) so complicated?" If all known bacteria contain a few thousand molecular species (coded by a few million base pairs in the DNA), it looks as if this might be the minimum-size genome. Yet all bacteria come to us today at the end of more than three (and perhaps more than four) billion years of evolution. How simple might the first organisms have been, and how simple might a microbe on Mars, Europa, or Titan be? Are we blinded by earth-style complexity when there can be much simpler life? This is why we linger at length over this seemingly trivial but obviously complicated matter. What indeed is the simplest assemblage of chemicals that could be considered alive? Perhaps the simplest earth life is among the most complicated of life-forms in the cosmos, and we earthling students of life are blind to that fact. The answer is not yet known. But there are at least a few people working on it, most in the fraternity of astrobiology. My sense is that there is a whole diversity of life simpler than the cells we deem the simplest. So here I propose the first heresy of many to come: that all around us are just such "simpler" well-known life-forms that we do not consider to be living. Not so. Viruses are alive.

The case for a living virus

Case in point, there are biological entities simpler than bacteria on Earth. Much simpler, in fact. Viruses, the scourge of humanity, the scourge (but perhaps the creator as well) of all life on Earth, clearly have a huge effect on life. But is a virus itself alive? This question has great importance in understanding the evolution of life on Earth as well as in classifying earth life, for if viruses are deemed alive, fundamental changes must be made to the classification of "life as we know it." Here I will make that case.

A virus is very small. Typical viruses are from 50 to 100 nm in diameter, where *nm* stands for "nanometer," or 10^{-9} meters. At this size many millions of viruses can fit on the head of a pin. They come in two general types: One group is enclosed in a shell of protein; the second in both a protein shell and a membranelike envelope. Within this covering is the most important part of the virus, its genome, made up of a nucleic acid component. It turns out that there is an amazing variety of genome types in viruses. In some there is DNA; in others only RNA. The number of genes also varies widely, with

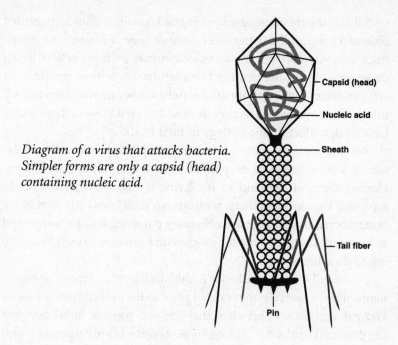

Diagram of a virus that attacks bacteria. Simpler forms are only a capsid (head) containing nucleic acid.

Capsid (head)

Nucleic acid

Sheath

Tail fiber

Pin

some having as few as 3 genes and others (such as smallpox) having more than 250. In fact, there is a huge variety of viruses, and if they were considered alive, they would be classified across a great taxonomic spread. But common wisdom treats them as nonliving.

The viruses that contain only RNA may be important in understanding questions surrounding the origin of life on earth. RNA viruses show that RNA by itself, in the absence of DNA, is capable of storing information and serving as a de facto DNA molecule. This finding is strong evidence that there may have been an RNA world before DNA and life, at least as we know it, originated. And there is an even more striking implication of the presence of RNA viruses. Since viruses are parasites, it may be argued that there may be as yet undiscovered RNA life-forms that are not viruses on our planet, with the RNA viruses as their parasites, as well as a proof of existence. Thus there may be simpler life than even a bacterium (but not as simple as a virus) living unknown among us.

All viruses are considered parasites. They are technically termed obligatory intracellular parasites because they are unable to reproduce without host cells. In most cases, viruses infiltrate cells of living organisms, hijack the protein-forming devices within the invaded cells, and start making more of themselves. They turn an invaded cell into a virus-producing factory. It is all too well known that viruses have a huge effect on the biology of their hosts.

We can use the various aforementioned definitions of life to decide if viruses are alive or not. The greatest argument against including them in the land of the living is that they need hosts to replicate; they are unable to replicate on their own. But it must be remembered that viruses are *obligatory* parasites, and parasites tend to undergo substantial morphological and genetic changes in adapting to their hosts.

Freeman Dyson considers it highly unlikely that viruses are remnants of primitive cells that existed prior to the evolution of life as we know it, and thus, in his view, they are not "missing links" between the dead and the living. Not everyone agrees with his argument, and as we will see in the next chapter, there is an enormous diversity of viruses just now being discovered, especially in the ocean, and many of these show some very peculiar and tantalizing traits. One group of RNA viruses studied by Alan Weiner and Nancy Maizels in 1987 were even described as living fossils of the RNA world! I will make this same argument, acknowledging here that Weiner and Maizels realized this many years ago.

That viruses may in fact be "alive" is also advocated by Martin Olomucki in his useful book *The Chemistry of Life*, in which he notes: "It is not always easy to draw a clear line between the largest viruses and the smallest primitive parasitic bacteria." One such is the infectious agent of parrot disease, or psittacosis, which was long thought to be a virus but which, under closer scrutiny, was identified as the primitive prokaryote *Chlamydia*. This very nasty bug is *very* small even for a microbe, 0.2 to 0.5 thousandths of a millimeter, and contains DNA, RNA, and a few enzymes. What it does not have, however, is any way to replicate itself outside its host. It is thus just like a virus, yet *Chlamydia* is definitely classified as living. By looking at a number of more complicated viruses, Olomucki concludes: "[Y]et, despite the poverty of its constitution, the virus cannot be considered merely a chemical substance; viruses are really and truly living objects." But if so, where do we put them in the classification of earth life? They are not Bacteria, or Archaea, and certainly not Eukarya, the three great domains of life (we revisit the concept of a domain in chapter 2 as well as the three domains of life on earth mentioned here). So what are they?

No one disputes that a virus is a parasite. So let us ask, are other parasites alive? Parasitism, which is essentially a highly evolved form of predation, is generally the result of a long evolutionary history. Parasites are not primitive creatures. But like our viruses, they have stages that do not seem fully alive. *Cryptosporidium* and *Giardia,* for instance, both well-known (and nasty!) parasites on humans and other mammals, have resting stages that are as dead as any virus outside its host. Without the hosts, the parasites mentioned earlier (and thousands of other species as well) will not live. Accepting these eukaryotic parasites, but then rejecting viruses as living, is not very scientific, in my opinion. If one group is, both are. So in this book, using the logic of those who study animal parasites, I will promote the heresy (to some anyway) that they are alive. These parasites are certainly alive, yet they cannot replicate outside hosts, just like viruses. And if we accept that viruses are alive, we must radically reassess the tree of life as it is currently accepted.

Are prions alive?

The existence of prions is also a test of the what is alive question. Prions are small and mysterious "organisms" that cause a number of truly hideous diseases, the most notorious being scrapie (in sheep) and mad cow disease, which can infect humans. When these diseases were first discovered, it was assumed that the infectious agent was a virus. Eventually the "agent" was discovered and named a prion, short for proteinaceous infective particle. Prions test negative for nucleic acids of any kind and thus have neither DNA nor RNA. It is conjectured that they are pure proteins, and there is now much evidence for this protein-only theory of prions. In fact, recombinant prions that are infective to mice have now been made. The mechanism of propagation is believed to involve the ability of the misfolded form of a normal cellular prion protein to cause a shape change in that protein, converting it from the normal shape (harmless) to a new shape that causes great harm to the host, if that is the correct word for the organism containing these "bad" prions. Prions are made/propagated in the cytoplasm in yeast and are thought to be made/propagated on the cell surface in mammals.

Replicate they do, when they invade nerve cells of their hosts and, like a virus, induce the host cells to start making lots of them. Like the presence of RNA viruses, the existence of prions has interesting implications. They may prove that protein alone can act as both software—coding information—and hardware. Do prions metabolize? Doubtful, since the protein-only theory seems to hold, and it is difficult to envision how a single protein could metabolize. Do they evolve? Very possibly, since the different strains are not equally efficient, and only one will propagate in a single host when a mixture is present, but each will propagate in a single host if introduced serially. Are they alive? Maybe. But like viruses, they cannot be placed into any standard taxonomy of life on Earth.

A heretical position

As we will see, one of the major definitions of life is that it is cellular. This is a point strongly made by biologists Lynn Margulis and Dorian Sagan in their book *What Is Life?* I disagree with this position. Here I take a page from musings by biologists Carl Woese and John Baross.

What if life can be an ecosystem as well as a cell, given the right kind of ecosystem? Early in earth history, in some primordial ooze, we may have had *thousands* of different kinds of protocells, naked genes, viruses, viruslike things, prionlike things, raw organic molecules, and on and on. By themselves none are alive. Taken altogether, the mishmash fulfills all requirements of life. This is my view too. If we landed on Titan, and a similar type of "organism," formed of myriad, dispersed, but chemically interacting components, were present, would we even recognize it as life? For that matter, if in some time machine we went back to the early earth, would we recognize life when it first started? Doubtful. Should we give a name to such life? Probably. But I'll be busy enough naming things in the pages to come that I'll pass on this one.

We might summarize the what is life question by suggesting that our view of life is enormously colored by our experience of being complex organisms in a world of complex organisms. Life can be simpler than the deluxe earth models, as evidenced by viruses. And it may well be that even viruses are very complex forms of life compared with varieties of life that we might find beyond Earth or hidden and unrecognized here on Earth. There is an enormous diversity of animate chemistry, at least theoretically. Ever since Darwin, it has been assumed that all our earth life is derived from a single common ancestor. But is this correct? What made Darwin think so, and how has that view changed with the revolutionary advancements we have made in microscopy in the minute chemical parsing of cells? These will be the subjects of the next chapter, where we go from defining life to looking at the nature of earth life. It used to be that these were one and the same. Once we can handily characterize what earth life is, we can then attack the problem of what earth life is *not*, life that is not earth life as we know it—like viruses, for instance—of the earth, but not earth life according to current biology. Viruses are thus aliens, our first of this book so far.

Chapter 2

What Is Earth Life?

This four-letter alphabet provides the basis for the common DNA language of all living organisms on Earth. They all share the same four-letter alphabet and DNA language, which is convincing evidence that we are all descended from one uniquely successful ancestor, whether that ancestor first appeared inside a comet or in Haldane's primeval soup or in Darwin's warm little pond.

—John Gribben

We humans have long and intuitively known that we each are the product of nature and nurture, a melding of our genes and the events in our lives and even the events in history before our lives. So too it must be for the life on any planet or moon. That has certainly been the case for earth life, which comes to us battered and melded by history. In the last chapter we looked at what life is. In a nutshell, life replicates, metabolizes, and evolves. So, in this chapter, let us establish what *earth* life is—that good old (really old) familiar stuff that we belong to—and ask how not only genes but also history have colored it. Seems like a simple enough prospect, with all the life there is around us to use as an example. But that is part of the problem: There is *so* much life on Earth and in such diversity of shape, habitat, and chemistry that coming up with a single definition requires some thought. Nevertheless, it's a pretty straightforward task. But there is a second and paradoxical aspect to this chapter's main task, one that has the faint whiff of irony. For all the thousands of pages written in pursuit of classifying the myriad aspects of life on Earth, there does not seem to be a formal published description of what earth life (or at least the kind of earth life that we belong to) *is*, so that we can know what it is *not*. It is just referred to as life, and since few, if any, of the army of taxonomists, those biologists preoccupied with classifying earth life, have ever considered life off the planet or had the temerity to question the ironclad doctrine delivered by Charles Darwin that all life on Earth comes from a single ancestor, somehow, no formal description of our kind of life has been made. There is not even a formalized taxonomic category, such as genus, species, or kingdom, that could be used to house such a definition. In this chapter we shall remedy that, to my surely everlasting infamy among classically trained biologists.

The construction of habitable planets in the solar system

Time is something that is integral to the story to come. Here and there I will have to refer to geological time, rather than the years in the past in which some event happened. Most of the events discussed here took place before the advent of animal life on Earth and hence at a time before common fossils. While the differentiation of rocks and time based on fossil content works quite well for the long interval of time since the rather sudden appearance in the rock record of larger fossils, the time before fossils—by far the longest interval of earth history—is more difficult to subdivide. This pre-Cambrian time is broken into three major divisions, named (from oldest to youngest), the Hadean, Archean, and Proterozoic eras. The Hadean was the time before life and any sort of abundant rock record. The Archean era began with the first appearance of life and a rock record but ended not with any biological event but with a series of physical changes to the earth. The succeeding Proterozoic was a time dominated by microbes, but near its end the first animals evolved, and the boundary between the Proterozoic and the succeeding Paleozoic is marked by the Cambrian explosion, when skeletonized animals appeared in large numbers for the first time. The Hadean, Archean, and Proterozoic eras are thus long intervals of time with few definable events.

So what was the planetary history that affected earth life? At the risk of departing from our main narrative, let us take a detour and look at how dear old Mother Earth came to be.

Life seems to have appeared on this planet somewhere between 4.1 and 3.8 billion years ago, somewhere near the end of the Hadean

Time (in millions of years ago)	Era	Events
65 to present	Cenozoic	First large mammals to end of terrestrial communities. End of ice age
250 to 65	Mesozoic	First dinosaurs to Cretaceous mass extinction
543 to 250	Paleozoic	First skeletonized animals to Permian mass extinction
2,500 to 543	Proterozoic	First eukaryotic cells to first skeletonized animals
3,800 to 2,500	Archean	First life to first eukaryotes
4,600 to 3,800	Hadean	Origin of earth to origin of life
Before 4,600	No name	Solar system (and planet Earth) forms

Table 1. *The Geological Time Scale*

or early in the Archean era—or some 0.5 to 0.7 billion years after the earth originated, but this is a window of time early in the earth's history when no fossils were preserved, thus obscuring our understanding of life's earliest incarnation. The oldest fossils that we find on the planet are from rocks about 3.6 million years of age and look identical to microbes still on Earth today. There may have been earlier types of life not now represented on Earth (or not yet found, but that is part of our story), but our present knowledge suggests that simple oval or spherical bacterial-like forms were the first to fossilize and may have been the shape of the first life on earth as well.

While life may be old, our planet is much older yet. Earth formed about 4.5 to 4.6 billion years ago from the coalescence of various size planetismals or small bodies of rock and frozen gases. For the first several hundred million years of its existence, a heavy bombardment of meteors continuously pelted the planet with lashing violence. Both the lavalike temperatures of the earth's forming surface as well as the energy released by the barrage of incoming meteors during this heavy bombardment phase would surely have created conditions inhospitable to life. The energy alone produced by this constant rain of gigantic comets and asteroids prior to about 4.4 billion years ago would have kept the earth's surface regions at temperatures sufficient to melt all surface rock and keep it in a molten state. There would have been no chance for water to form as a liquid on the surface. Clearly there would have been no chance for life to either form or survive on the planet's surface.

The new planet began to change rapidly soon after its initial coalescence. About 4.5 billion years ago the earth began to differentiate into the different layers. A lower-density region called the mantle surrounded the innermost region, a core composed largely of iron and nickel. A thin, rapidly hardening crust of still-lesser-density rock formed over the mantle, while a thick, roiling atmosphere of steam and carbon dioxide filled the skies. In spite of its being waterless on its surface, water was associated with the planet. Great volumes of water would have been locked up in the interior of the earth and would have been present in the atmosphere as steam. As lighter elements bubbled upward and heavier ones sank, water and other volatile compounds were expelled from within the earth and added to the atmosphere. The constant bombardment by giant comets and asteroids lasted more than a half billion years and finally began to diminish around 3.8 billion years ago as the majority of meteors were gravitationally pulled into the various planets and moons of our solar system. During the heaviest impact period the steady bombardment would have cratered the planet in the same manner as the moon. Yet the comets and asteroids raining in from space delivered an important cargo with many of these titanic impacts. Some astronomers believe that much or most of the water now on our planet's surface arrived largely through the incoming comets; others think that only a minority of the earth's water arrived in this fashion. There is still no resolution to this question, although some new measurements of comets suggest that they were not the main source of our water, the elixir of life on Earth and perhaps life everywhere. And the comets brought other components for life, including carbon compounds. Instead of Earth going to the chemical supply store for the ingredients of life, the store came to it.

Comets are made up mainly of volatiles, such as water and frozen carbon dioxide, and there is no doubt that a good many of them hit the early earth. Study of carbonaceous asteroids and comets shows that many carbon compounds rained onto the surface of the earth in the earliest ages of the solar system. And not just on Earth. Looking at the moon and Mars reveals the ubiquity and abundance of craters. Enormous numbers of comets rained down onto all the planets and moons, bringing the stuff of life to every moon and planet in the solar system. For some, such as Mercury and the outer gas giants, these gifts of life's components were entirely superfluous. But for Earth and perhaps Venus, Mars, Europa, and Titan, the five bodies with life (or the best chance of having had it or still having it), the rain of comets was probably instrumental in whatever organic evolution that occurred.

Prior to 4.4 billion years ago these cargoes of water slamming into the earth at least would have turned instantly to steam because of the high temperatures on the surface. Perhaps the same happened on Venus and Mars. But Earth and its inner and outer sister planets gradually began to cool as their heat dissipated into space. As early as 4.4 billion years ago surface temperatures on these three planets might have dropped to below 100°C, and for the first time liquid water would have condensed from steam onto our planet's surface, in the process successively forming ponds, then lakes, and seas, and finally a planet-girdling ocean. It is thought that the same happened on Mars and Venus. If seen from space, there may have been *three* blue jewels orbiting the sun back then instead of the single blue planet of the present day, our water-drenched Earth. A watcher from Earth back then would have remarked on the beauty of blue Venus and blue Mars. The study of ancient sedimentation suggests that by slightly less than 3.9 billion years ago, the amount of oceanic water on Earth may have approached or attained its present-day value, along with that on Mars and Venus. This time might have been the period of the most widespread habitability in our solar system. But they were not tranquil oceans on these planets, or oceans even remotely familiar to those of today.

We have only to look at the moon to be reminded of how peppered the earth and its oceans were during the period of heavy impact, between 4.4 and 3.9 billion years ago. Each successive large-impact event (caused by comets as large as 500 kilometers in diameter) would have partially or even completely vaporized the oceans. Imagine the scene as viewed from outer space: the fall of the large comet or asteroid, the flash of energy, and the vaporization of the earth's planet-covering ocean, to be replaced by a planet-smothering cloud of water vapor and rock-filled steam heated (at least for some decades or centuries) far above the boiling point of liquid water. It is difficult to conceive of life, whatever its forms, surviving anywhere on the planet during such times, unless that survival occurred deep underground.

NASA scientists have completed mathematical models of such ocean-evaporating impact events. The collision of a 500-kilometer-diameter body with the earth results in a cataclysm almost unimaginable. Huge regions of the earth's rocky surface is vaporized, creating a cloud of superheated "rock gas," or vapors, several thousand degrees in temperature. It is these vapors in the atmosphere that cause the entire ocean to evaporate into steam. Cooling by radiation into space would take place, but a new ocean would not rain out for at least several thousand years after the event. Much of the revolutionary detective work behind these conclusions was described in 1989 by Stanford University scientist Norman Sleep, who realized that impacts of such large asteroids or comets could evaporate a ten-thousand-foot-deep ocean, sterilizing the surface of the earth in the process.

It is ironic that the comets may have brought some of the earth's life-giving liquid water, the necessary prerequisite of life, and then taken that gift away for a time with each successive large-impact event. Yet it is not only water that these comets may have brought. They played a role in determining the chemical evolution of the earth's crust. And they perhaps brought another ingredient in the mix we call life: organic molecules or even life itself onto our planet's surface for the first time.

About 3.8 billion years ago, at the end of the period of heavy bombardment, our world would surely still appear alien to us. Even though the worst barrage of meteor impacts would have passed, there still would have been a much higher frequency of these violent collisions than in more recent times. The length of the day was far different, being less than ten hours long, because the earth was rotating far faster than it does today. The sun would appear much dimmer, perhaps a red orb of little heat, for it not only was burning with far less energy than today, but it had to shine through a poisonous, riled atmosphere composed of billowing carbon dioxide, hydrogen sulfide, steam, and methane. In such an environment we would have had to wear spacesuits of some sort, for only tiny traces of oxygen were present. The sky itself would probably have been orange to brick red, and the seas, which surely covered most of the earth's surface (save for a few scattered low islands) would have been muddy brown and clogged with sediment. Yet perhaps the greatest surprise to us would be the utter absence of life. No trees, no shrubs, no seaweed or floating plankton in the sea; it would have seemed a sterile world. Somehow the fact that we have not yet detected life on Mars seems consistent with its satellite images. A waterless world fits our picture of a lifeless world. But even when the young earth was covered with water, it was still devoid of life. However, it was not for long.

For Venus and Mars at this time a different history was playing out. Mars, the wet planet, was dying and drying out. On both planets their blue oceans were lost to space. Gradually, Mars became a cold desert and Venus an unbelievably hot greenhouse. From that time on, the surface of neither planet would have been fit for life, as we know it anyway. But on both places there remained potential refuges, as we will see later.

Most scientists are confident that life on Earth had already arisen 3.8 to 3.9 billion years ago, at about the time when the heavy bombardment was coming to an end. Because of the suspected similarity in conditions on Mars and Venus at the time, there may have been life on these two worlds as well. Indeed, the very violence of the impacts at that time may have seeded each of the planets with life from one of them, a process called panspermia that we will revisit later. We earthlings might really be Martians, or Venusians, or their life might have been from Earth. Or perhaps life arose three separate times. We have no shortage of hypotheses to be disproved.

The evidence indicative of earth life's appearance is not the presence of fossils, but of isotopic signatures of life extracted from rocks of that age in Greenland. The oldest rocks on Earth that have been successfully dated using radiometric-dating techniques are mineral grains of zircon, yielding ages of about four billion years. The Greenland rocks (from a locality named Isua) are thus only slightly younger. The Isua rock assemblages include sedimentary (layered rocks) and volcanic rocks and have yielded a most striking discovery. They contain isotopes of carbon, life's most diagnostic elemental signature, suggesting that they were formed in the presence of life. The isotopic residue in the Isua rocks is an excess of the isotope carbon-12 as opposed to carbon-13. A surplus of carbon-12 is found today in the presence of photosynthesizing plants, since all living organisms show an enzymatic preference for "light" carbon. The inference is that if early life existed at Isua, it may have used photosynthesis for its energy sources. But there is no fossil evidence that life existed this long ago, only this enigmatic and provocative surplus of a carbon isotope that in the present world is a sign of life's presence. If the excess of light carbon isotope is indeed a reliable indication that ancient life existed at Isua and perhaps elsewhere on the earth as early as 3.8 billion years ago, it leads us to a striking conclusion: Life seems to have appeared simultaneously with the cessation of the heavy bombardment period. As soon as the rain of asteroids ceased, and surface temperatures on Earth fell below the boiling point of water, life seems to have appeared. But how?

There are still more questions than answers about life's origins on earth, and obviously nothing but questions concerning potential life on Mars and Venus. Yet the sophistication of the questions now being addressed by a legion of interested scientists tells us that we are well along in the investigation. Among the most pressing of these questions: Did the origin of life occur in only a single or in several settings? Did the key chemical components, the building blocks, come from different environments to be assembled in one place? Was life's origin deterministic—i.e., could there be different environmental conditions producing the same molecule of life, the familiar DNA? Were the individual stages in the origin of life (such as the formation of amino acids, then nucleic acids, then cells) dependent on long-term changes in the earth's environment? Did the origin of life change the environment such that life could never originate again? At what stage did evolution take over to guide the development of life? Perhaps most interesting of all, can we infer the nature of the settings of life's origin from the study of extant organisms, creatures living on Earth today?

But what really *is* earth life?

No one would dispute that planet Earth is besotted with life. From that long-ago time when earth life first formed, we have gone from some first cell to millions of species. From the deepest ocean depths to the highest mountain crags, it is difficult to find any habitat that does not at least harbor a smattering of microbes, and when we reach the richness of the African plain, or tropical rain forest, or coral reef habitat, the cornucopia that we call life is everywhere apparent. It is this very richness that has caused no end of dispute and work for those biologists concerned with classifying it all. But even with all this diversity, all life yet discovered shows a unifying characteristic: It all contains DNA. So perhaps this is how we should define earth life.

Deoxyribosenucleicacid, or DNA, is something that we encountered first in biology class and then constantly in the news. If nothing else, the infamous O. J. Simpson trial brought this peculiar molecule to the fore of our consciousness. Detecting the identities of criminals, establishing paternity, following the endless crime

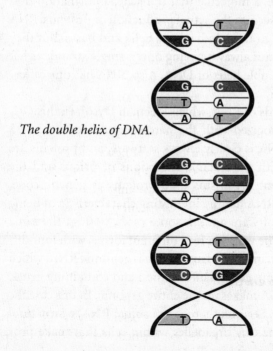

The double helix of DNA.

scene forensics of the scandalous trial de jour, DNA as a concept cannot be escaped. Yet it is complex, and its actions even more so, and since so much of this book deals with it in one variation or another, it might be profitable to describe it in detail.

Composed of two backbones (the famous double helix described by its discoverers, James Watson and Francis Crick), this complex molecule is the information storage system of life itself, the "software" that runs all of earth life's hardware. These two spirals are bound together by a series of projections, like steps on ladder, made up of the distinctive DNA bases, or base pairs: adenine, cytosine, guanine, and thymine. The term *base pair* comes from the fact that the bases always join up: Cytosine always pairs with guanine, and thymine always joins with adenine. The order of base pairs supplies the language of life; these are the genes that code for all information about a particular life-form.

If DNA is the information carrier, a single-stranded variant called RNA is its slave, a molecule that translates information into action—or in life's case, into the actual production of proteins. RNA molecules are similar to DNA in having a helix and bases. But they differ in usually (but not always) having only a single strand, or helix, rather than the double helix of DNA. Also, RNA has one different base from DNA.

There are four kinds of RNA, which Freeman Dyson (in his *Origins of Life*) has analogized with the hardware and software of a computer system. DNA is clearly always software, and proteins are usually hardware (with the notable exceptions of prions and the possibility that ancient organisms used proteins as genetic code, making it software). RNA has the interesting characteristic of being either hardware or software and, in some cases, both at the same time. RNA occurs in the world in four different forms, with four different functions. First, in some viruses there is genomic RNA, which acts like DNA in storing genetic information and containing genes. In the AIDS virus, RNA makes up the entire genome. In this case the RNA acts as software. Second, there is ribosomal RNA, a structural part of ribosomes, the tiny organelles within cells that make proteins. This is a case of RNA clearly acting as software. Third, there is transfer RNA, the hod carrier that takes amino acids to ribosomes for protein synthesis, and as a material conveyor it is hardware. Finally, there is the most interesting of all the RNAs, messenger RNA, which conveys instructions to the ribosome from the genomic DNA. In this it acts as software, but it has been shown that it can also act as a catalyst both for protein formation and for its own splitting and spicing and thus acts as software *and* hardware at the same time.

Understanding RNA, its use and evolution, is key to understanding life on Earth, and perhaps not only earth life but other types of life as well. In earth life DNA makes RNA, which makes proteins. This is known as the central dogma and was first defined by Francis Crick (who later admitted that he regretted using the term *dogma* in this definition). But as we will see, RNA might have preceded DNA during earth life's origin. Most RNA is used as a messenger, sent from DNA to the site of protein formation within a cell, where the specific RNA gives the information necessary to synthesize a particular protein. To do this, a double-stranded DNA partially unwinds, and a single-stranded RNA forms and keys into the base pair sequence on the now-exposed DNA molecule. This new RNA stand matches with the base pairs of the DNA and in so doing encodes information about the protein necessary to be built. This brings us to the subject of genes.

DNA provided the answers to many of the mysteries of genetics, answering the question, once and for all, about what a gene is. Watson and Crick made the great discovery, one that launched an enormous revolution in biology, and it was announced in a paper in the journal *Nature* that was a single page long. Their finding was actually a model, not an experimental result, but the model had enormous predictive power. It became clear that a gene is made of DNA and that one gene makes one protein. Watson and Crick proposed that one-half of the DNA ladder serves as a template for re-creating the other half during replication. Each gene is a discrete sequence of DNA nucleotides, with each "word" in the genetic code being three letters long.

How does a gene specify the production of an enzyme? It was Crick who suggested that the sequence of bases is a code, the so-called genetic code, that somehow provides information for the formation of proteins, one amino acid at a time. The information coded has to be read (transcribed) and then translated into proteins. That is where RNA comes in. Life as we know it uses twenty amino acids. Not nineteen. Not twenty-one. *And always the same twenty!* If we suddenly found (or made!) life that used a twenty-first amino acid, for instance, this would be a good reason to rejoice in the discovery of alien life. There is a transfer RNA molecule specific for each of the anointed twenty of the amino acid clan. Once alerted by Chief DNA that a particular amino acid from the twenty is needed for a specific protein to be built, our transfer RNA goes out into the cytoplasm of the cell interior, scavenging for the particular amino acid that it alone can carry. Once the amino acid is found, this transfer RNA then heads to the ribosome with its burden.

The code is elegant and can be analogized to Morse code, itself just a system of dots and dashes that is able to string together long and complex messages. Crick realized that the different combinations of bases lined up on the DNA molecule could specify each of the twenty amino acids used by life on Earth. But actually making the proteins took place in the small spherical bodies within the ribosomes. Therefore, some link had to be made between the DNA and the protein formation centers. This is the job of messenger RNA molecules. Thus DNA codes for RNA, which codes for proteins. This, then, the central dogma of molecular biology, may also be called a central characteristic of earth life.

How to define earth life

We are starting to compile a large number of specific characteristics that our kind of life uses to stay alive. Let us look at the genes of earth life in more detail, so as to understand how they might differ in nonearth life. First of all, genes are the blueprints necessary to make earth life's major structural and chemical partner, proteins. Proteins perform the various functions of the cell. A protein's action is determined both by its chemical constituents and by its shape. Proteins become folded in highly complicated topographies, and often their final three-dimensional shape determines their actions.

So, how does DNA specify a particular protein? A typical protein might be made up of a hundred to more than five hundred amino acids, and thus its gene, the sequence of nucleotides coding for the protein on the DNA strand (since the string of amino acids that make up the protein are coded on the DNA strand), will be composed of a hundred to five hundred or more sets of "steps" on the DNA ladder. These are arranged in linear order along the DNA strand, like letters in a sentence. And like a sentence, there will be spaces and punctuation as well (like *stop*!). The RNA slaves grab these and take them to a ribosome, where the actual protein is constructed.

So, earth life has DNA, and RNA, has a specific code, and uses tiny structures in the cell called ribosomes to make another characteristic of earth life, proteins.

The code itself is important to look at, for it is one area that could be changed to produce alien life or at least form a DNA life unlike that on Earth. The fact that all our bodies are made up of proteins constructed from twenty different amino acids, but always the same twenty, is itself a characteristic of earth life. Again, using different, more, or fewer amino acids would certainly seem to qualify a life-form as being unearthlike.

This information flow goes only one way: from DNA to RNA. The poor RNAs have no say in any of this: go here, build that, bossed forever from above by DNA. All the proteins being built by the ribosomes, at the direction of the RNAs (themselves slaves to the DNA), do one of two things: They build a structure, or more commonly, they function as enzymes that catalyze a chemical reaction in the cell itself important for maintaining life function, such as metabolism.

Our description has gotten more complex. We need to incorporate the code that is used and the twenty amino acids that we are made of. Now we have more to play with: We have a specific information-carrying molecule (DNA) that is found in a structure (a chromosome) that (using RNA and ribosomes) works to produce a slew of proteins, all made up of twenty, only twenty, specific amino acids (these can be found in any biology text), using a particular code of nucleotides to specify an individual amino acid.

This much information is no longer a rough sketch of a suspect in a police report but a fingerprint. Is this fingerprint unique to life on Earth or even to all life on Earth?

A new classification for life

What else can be used to diagnose (the formal term used by taxonomists) this taxon (any one of the formal units of biological classification) that constitutes the common form of earth life? Again we have to hedge, for as we shall see in chapter 5, there are other forms of life, other taxa, distinct from those of our still-unnamed life as we know it, that have existed in the past and may exist now on the planet, as well as artificially produced life that does not meet this diagnosis. We also have to hedge about the taxon, for without giving away too much of the surprise, not only is there not yet a taxon for life as we know it, but there is not even a taxonomic category yet defined that would include all of this life as we know it. So let us look at what else there might be that categorizes this group of life and then look at the currently accepted tree of life and as well at how such trees are constructed. That done, we can start tree building.

Having DNA is obviously not all there is to life. We need a wall (membrane) to enclose our cell and a solvent to fill it with. Both the specific wall or membrane structure and the specific solvent are also features that we can use to identify common earth life. In a recent insightful essay in *Current Affairs,* biochemist Steven Benner (a member of my University of Washington NASA Astrobiology Institute team who appears again later) and two colleagues have described our familiar life in even more basic chemical terms. They see life as needing isolation of some sort, within a membrane or bounding domain that is chemically produced. But they also point out that isolation does not necessarily require putting guts in a 3-D box. It can be achieved on a two-dimensional surface, and this understanding has led to new ideas about the interaction of organic molecules and mineral surfaces that have greatly expanded our view of how life came about and how it might be on non-Terran habitats.

Benner et al. also suggest that a requirement of life is some sort of scaffolding, for building blocks of our life structure and for holding biomolecules in correct orientation so as to allow chemical processes of life. Our earth life uses carbon as the scaffolding element, but silicon could be used as well if there were side branches on which carbon compounds could bond. One such material is called oligosilane, which is made up of silicon atoms that are bonded together in chains (but that also have side chains that are not composed of silicon). These have the potential to make membranes that would be very different from the cell wall membranes used by our familiar carbon life. We can add to our definition that our familiar earth life has a plasma membrane, composed of a cell wall composed of material known as phospholipid bilayer and membrane protein and a specific solvent, water. Now we have a description of this kind of life. Let's give it an informal name, so we can dispense with the very clumsy life as we know it, earth life, or DNA life that we have been using to this point. There could be any name, but following old practice, let us utilize the Latin for earth (*terra*) with the suffix used to denote life (*oa*, meaning "life") and combine this into the name Terroa. It could be Terraoa, but that seems a little too Polynesian or denoting terrorists of some sort. So: Terroa. Earth life. The life as we know it kind of earth life anyway.

There are lots of Terroans around us, and there have been for at least 3.7 billion years. But not always. Sometime, in the deep past of our planet—perhaps just more than 3.7 billion years ago, perhaps more than 4 billion years ago—there was life but not Terroan life on Earth. This was before some first cell combined all the DNA and

RNA apparatus, membranes, water, enzymes made from the twenty amino acids, and so forth—the whole shebang that we DNA life-forms use—to form the very first Terroan, living in a fashion that might not seem strange to those who know the ecology of microbes in the present day. There was some first cell of some first Terroan species, the only one of its kind. There is no fossil that memorializes it. Nevertheless there must have been some first example of what we now call earth life. It and the rest of its species, for there were more than one of these little urchins, even have a name: LUCA, for last universal common ancestor.

Other terms have been used to describe this ancient parent of us Terroans. The pioneering microbiologist Carl Woese first called it the progenote. But the meaning of *progenote* seems to denote a particularly primitive ancestor, one much simpler than actual cells. This would not be the first species that we could call earth life, but it is something leading up to it. LUCA is viewed as a microbial common ancestor, resembling either Bacteria or Archaea.

Was our first Terroan the first life? Was it thus lonely? Definitely not, on both counts. Our LUCA did not pop out of the firmament fully formed. It had ancestors. And they were not Terroans and thus need their own name. In fact they were aliens, definitely life as we do *not* know it. But that is part of our story to be taken up in the next chapter. At the moment let us contemplate LUCA, our first Terroan, a unique kind of species on Earth, but probably not a unique species of life. LUCA was surely surrounded by a huge diversity of life. There would have been other true cellular life, like LUCA, as well as a whole zoo-full of other life-forms and unloving but organic molecules, a real bouillabaisse of early life on Earth in the primordial soup (if there was a soup, which some now disbelieve).

In all this surely great slew of life-forms, LUCA was indeed unique but probably not for long. The earth back then, as now, was a place of many habitats and with many ways that a resourceful form of life (like our little LUCA, the most modern life on the planet, a veritable prodigy on that ancient earth) could acquire energy and material for growth. Natural selection would have stoked the fires of the speciation process, and where there had been but a single species with DNA and proteins as we use them, there now would have been two, and soon after, many, many more. Meanwhile, as our sharp little Terroans ran rings around the poor RNA life and other losers, wholesale extinction of early life-forms surely took place. The burning of the great Roman library in Alexandria was as nothing compared with the loss of information during this first great mass extinction on earth, as Terroans co-opted the planet's resources in ruthless Darwinian fashion, and untold genomes of pre-Terroan life blinked out of existence. The standard story is that at the end of this evolutionary winnowing, only the Terroans were left as life on Earth. But I dispute that. Like the coelacanth fish, a few of the more ancient kinds of life hung on. Indeed they are among us now.

Today we define species as composed of individual organisms that are capable of interbreeding successfully. The species is also the basic unit in life's classification, the method that biologists use to organize the enormous diversity of life on Earth. Because the evolutionary process causes one or more species to arise from other coexisting or preceding species, many share ancestors. Blocks of species are thus united by common heritage, just as the siblings of a family share a set of parents. These groups of related species are called higher taxa.

The Swedish naturalist Carl Linnaeus developed the modern way of describing organisms and of organizing both the species and the higher taxa. In 1758, under the title *Systema naturae*, Linnaeus published one of the great revolutionary works of science. He swept aside the old ways of naming and classifying organisms and heralded in an era that continues to this day. Very few other scientific advances from the middle of the eighteenth century have worn so well. Linnaeus proposed a binomial system for naming organisms, replacing the old system of a single common name for every living creature. Each species has two names, its genus and species names. To avoid further confusion, the name of the scientist formalizing the name is appended at the end along with the date on which the species was described. Linnaeus also grouped blocks of species into higher categories, on the basis of degree of similarity. In so doing, he erected the major taxonomic categories.

We live on a planet with millions of species. If they just appeared through some sort of divine creation, we would expect a lot of unrelated forms. But anyone can see that species form groups: dogs and wolves, house cats and lions, zebra fish and sharks, and on and on. We now know why there are such groups: Forms related through an evolutionary lineage almost always resemble one another more than they do other evolutionary lineages. Since Linnaeus, biologists have recognized that species can be grouped into hierarchical assemblages, but they have not always understood why. Charles Darwin, and his nifty theory of evolution, answered that particular why. Darwin referred to a classification based on evolutionary history as a natural system and explained why the Linnaean system had been so successful for categorizing animals and plants: They had been placed into the hierarchical groups of the Linnaean classification on the basis of similarities, and because these similarities reflected the evolutionary closeness of the respective species, the classification tended to reflect their historical relationships. Thus, the various taxonomic categories—the families, orders, and so on—could be understood as nested, or hierarchical, models of evolution. Lines of descent link these units: All species placed in any higher category share an ancestor. Species are grouped into genera, genera into families, families into orders, orders into classes, classes into phyla, and phyla into kingdoms. The kingdoms, until recently, were the highest level.

The methodology for illustrating how the various genealogies unfolded through time was the construction of what are known as phylogenetic trees, and because of the importance that they play later, the method should be described in some detail.

A tree is a very useful analogy for understanding evolution. A tree starts from a seed, grows roots (down) and a trunk (up), and then builds an ever more anastomozing series of branches out of the trunk. In evolutionary trees the branching pattern (often called branching order) shows the genealogy of the organisms, indicating which species share more common ancestries than others. While trees help understand the lineages of things, they are poor at illustrating the relative taxonomic level of things; there is no a priori reason that one part of our tree is a kingdom, and one a class. Assigning these categories is subjective.

Linnaeus and Darwin never foresaw that humans would one day head into space—or build alien life in a test tube—and therefore saw no need for any taxonomic category higher than kingdom. The earliest practitioners of this system first recognized only two kingdoms, animals and plants, and the system worked superbly for these larger creatures. But once microbial life became accessible to biologists with their ever more sophisticated microscopes, classification became more difficult, and the number of recognized kingdoms had to increase. Bacteria appear as three simple morphologies—rods, balls, and spirals—and hence, with so little morphology to deal with, were fairly impervious to classification based on morphology. They were relegated to their own kingdom, the Monera. As biologists better understood various plantlike groups, they increased the number of kingdoms to five—the animals, plants, fungi, protozoa, and bacteria—and this classification held until the late twentieth century. But with the advent of a new system of classifying organisms, using a sophisticated method of comparing genetic codes among various microbes (and other organisms as well), a whole new view of things, a true revolution in our understanding of life's order, appeared, using genotype instead of phenotype to track evolution. The revolution was possible because of the new and powerful methods of decoding genetic sequences and codes that became standard practice for evolutionists from the late 1960s into the 1970s. The methods, ever more powerful, are still very much used, and in a way they have made all other ways of classifying obsolete, unfortunately having the effect of limiting progress in thinking about life that is not Terroan since the dominant method relies on looking at the RNA in small organelles. But heresy! What if there was a life-form on Earth—or Mars, for example—that did not have DNA? I believe that we have reached that impasse. I think that I can show that such life currently lives on Earth. How could we classify such life? *Using current procedures, we cannot!*

So what is the party line? One of the now-standard techniques for comparing microbes at the molecular level, called DNA-DNA hybridization, takes DNA from one species of microbe and mixes it with the DNA from a second species. The similarity of the DNAs (hence the similarity of the two species) is reflected in the extent to which strands of DNA from one organism anneal with strands from the other. The problem with this method is that it works best for closely related species but is much less useful in distantly related species. In such cases, a much more useful method is to study phylogenies (the actual evolutionary pathways) by comparing the similarity and differences of the molecules making up a specific gene or protein that is common to both organisms. To be useful, the target molecule must be large enough to allow comparisons. One of the most useful of such molecules, found in all Terroans, is ribosomal RNA (the transfer molecule that DNA sends off to instruct the ribosomes in protein formation). By comparing the observed RNA sequences (or those of any other appropriate molecule), one can estimate both the historical branching order of the species and the total amount of sequence change.

While many workers in the early 1970s gradually began to use this new method, it was the microbiologist Carl Woese who first recognized the full potential of RNA sequences as a measure of phylogenetic relatedness—at least among Terroans. He began to compare the sequences from many different microbes and thereby initiated a revolution.

At first it would seem unlikely that the gene sequences preserved in still-living organisms could yield *any* sort of accurate key to the past, especially one of such antiquity. After all, the sequencing effort by geneticists is an attempt to unravel life's first diversification, which took place more than three billion years ago. Yet at least in some molecules, evolutionary change has been exceedingly slow. It was Woese (among others) who found that the most convenient places to study rates of evolutionary change within cells come from small subunits of RNA extracted from ribosomes; these have been the Rosetta stone, which gave a new view of Terroan evolution.

Woese was especially interested in a group of microbes called Archaeans. They had long been overlooked because they closely resemble bacteria. But once he was able to analyze their DNA by comparing their RNA sequences, it became clear that these tiny cells were as different from bacteria in genotype as bacteria are from the most primitive stocks of protozoa. This represented a huge dilemma: These differences were even greater than those found when Woese compared the RNA from the various kingdoms (animals, plants, and bacteria). Woese had discovered that the Archaeans were not even a separate kingdom; the differences were more profound. In an act of great intellectual bravery (one that had some very practical career enhancement benefits, as well as insuring his scientific immortality), Woese proposed, in 1976, that an entirely new category of life had to be erected, one that was above the level of a kingdom. He called it a domain. Using this new category, he proposed a whole new view of Terroan phylogeny.

The analysis of molecular sequences derived from living organisms, by Woese and others, thus provides a rough "map" of life's evolution as well as classifies it into groups. If portrayed in graphical form, this map becomes a "tree" like that mentioned earlier. The greater the numbers of differences between the genes, the more evolutionarily separated are the groups. It was this technique that showed the existence of the three fundamental groupings of organisms on Earth, groupings even more fundamental than the kingdoms, the domains Archaea, Bacteria, and Eukarya, and "showed" (or at least he thought that it did) that these three are the most ancient and basal branches of the tree of life still present on the planet. This analysis also showed that Bacteria and Archaea are distinct, even though both share some similarities, such as a cell without an internal nucleus. By the early 1980s the five kingdoms had became spread over the three domains: the Archaea, Bacteria, and the new category called the Eukarya, which included the old groups' plants, animals, protists, and fungi. The tree of the domains (and some of the kingdoms) is shown here, and it has come to be known as the tree of life.

The tree of life is really a model of life's evolution into the major categories of existing organisms. Various studies that compared gene sequences in various taxa gave a theoretical map of the evolutionary history of early life on Earth that first began to appear at about the same time as the discovery of the hydrothermal vents. According to these studies, there is nothing more "primitive" still living on Earth than Archaeans. They seem to show more characteristics and genes with the supposed primordial organism (the hypothesized common ancestor of all life) than any other living organism on Earth.

Much of this work contradicted long-held beliefs about the phylogeny of earth life with DNA. It showed that the division between the Bacteria and the Archaea is extremely ancient. Another surprise is the discovery that the Eukarya, the group from which higher animals and plants ultimately arose, is also extremely ancient. But by far the most intriguing result is that the most ancient of Archaeans and Bacteria are heat-loving microbes that are described as extremophiles, life that loves the extreme. In this case, the extreme they love

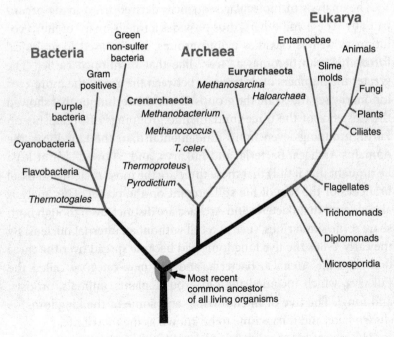

Traditional tree of earth life

is heat. These are just the types of microbes we find in extreme environments on Earth today. The discovery indicates that the earliest life on Earth was some sort of extremophile, suggesting that life may have first arisen on Earth under conditions of high temperature and pressure, either underwater or deep in the earth's crust, rather than in Darwin's pond, whose organisms would not have needed genes that supported life at high temperatures.

A new view of the tree of life

The tree of life is a hypothesis of evolutionary pathways as well as a means of classifying life. One of the key controversies has been the placement of the "root" of the tree, the point that tells us where the most primitive organisms lie. As the tree of life is composed of three domains, the Archaea, Bacteria, and Eukarya, it is a fair question to ask: Which came first? The currently accepted view is that the Archaeans came first and then gave rise to Bacteria and finally to Eukaryans. Lateral gene transfer, in which packets of genes jumped from group to group, was integral in this early evolution.

Over the last several pages I have been informally using the term *Terroan* for life as we know it. Earth life can be defined as follows: "Heredity is preserved in nucleic acids, proteins are made in ribosomes, the same set of amino acids is used to construct proteins, energy is stored in ATP, and an almost identical genetic code is used." Here I propose to formalize this concept by placing earth life within a wholly new category of life, which sits above domain. There are now sound reasons to justify this rather bold act. And let me do this here, in this book, making it at one and the same time a science book for the public and a science book for the scientists.

Here I define a new category and place a new group within it.

Name of new taxonomic category: **Dominion**
Definition. A dominion is a taxonomic category that is above the level of domain and is thus composed of domains.

Now I will formally name a dominion. The publication of this book formalizes it, according the rules and conventions of the Code of Taxonomic Practice.

> **Dominion Terroa, Ward, 2005**
> (*From the Latin,* earth life).
> **Definition:** *Life containing two-stranded DNA and utilizing it as its genome/information storage molecule; proteins made up of twenty amino acids (for specific amino acids), which are coded for by three-letter nucleotide sequences (for specific code); with a phospholipid membrane and water used as solvent in the cell.*
>
> At this time the Dominion Terroa contains three domains: *Archaea, Bacteria, and Eukarya.*

This done, we can now think about life that does not belong on this tree. And we can go looking for the non-Terroans. It turns out that we do not have to look very far. In just about any sneeze we fill whole rooms with aliens, if we define any life that is not currently on the Terroan tree of life as an alien. In chapter 1 I made the case that viruses are alive. If that is true, they have to be put on our tree of life somewhere. But where? To deal with them, we have to look at them in more detail.

What are viruses?

As we saw in chapter 1, a virus is a small bit of either DNA or RNA (its genome) packaged in a protein coat, known as the virion. To make a virus, we thus need nucleic acid of one kind or another and protein. The nucleic acid has to have some pretty complex instructions: how to break into a cell and then how to hijack its manufacturing machinery to make new viruses. Since both these products are difficult to make, the formation of the first virus is itself a perplexing problem in biosynthesis. But for a very long time, in spite of their obvious and interesting properties of parasitism, viruses were not fit subjects of study for any "frontline" biologist unless the biologist in question was interested in disease. Viruses were of no interest for their own sake. The viruses have been very clever at staying out of science's way. Because they contain no ribosomal RNA, the great tool pioneered and used by Woese and others to construct the tree of life (which is based on the differences in the sequences found in ribosomal RNA), viruses could not be compared with cellular life. Indeed, the prejudice that viruses were not alive only added to their neglect. Thus some very interesting problems, such as their origin, diversity, and ubiquity, have been overlooked. But recently this trend has changed, and viruses are getting new scrutiny. Amazing things are being discovered; some are very relevant to understanding the nature of life as we know it, as well as life as we don't. But first, let's look at viruses in general, and especially the new information about their ubiquity and diversity.

How many viruses?

The first thing that is apparent about viruses is that there are lots of them out there, and lots of different kinds. First, the how many. New research shows that viruses are almost unbelievably abundant, especially in the sea. While many microbiologists, on the basis of their realization of the staggering number of microbes found in almost any substrate or medium with any liquid at all (as well as some without, such as ice and rock), have rightly pointed out that we are in an age of bacteria, it has been estimated that there are from *one to two orders of magnitude more viruses* than there are bacteria. This is a staggering number. If we could make all the water disappear in the sea, for instance, we would see this ghost world entirely forged of viruses. They are found from the bottom of the sea (and well below the bottom of the sea) to the highest regions of the atmosphere and in every kind of life that has been examined. In one milliliter of seawater (about the volume of the tip of your thumb) there can be ten million viruses. As virologist Dennis Bamford has so unsettlingly described, "cellular life is bathing in a virtual sea of viruses," and he finishes this sentence with an even more chilling yet thought-provoking phrase:

"possibly creating the highest selective pressure they [we] encounter." What? Viruses affect evolution or perhaps channel evolution? There is an enormous importance in this last pronouncement, if true. That viruses might be of more importance in terms of adaptive pressure and natural selection than any other factor encountered by a cellular organism—more than food, habitat, predators, competition, and finding mates—seems at first glance ludicrous. Viruses, it turns out, partially compose all other life-forms. They also pervade the external biosphere and are capable of entering living cells—all living cells, it seems, with varying ease. It is as if they had the key to the tiny fortresses or castles that we call cells. They apparently have a master key or know the back-door tunnels that have been long forgotten by the cell's defenses whose job it is to guard against invasion by those that would enter and plunder. It is as if viruses were around while the first castles were being built or, more explicitly, as if among the original builders. There may be enormous implications to this that relate to the origin of life. What if viruses were as ubiquitous 3.7 billion years ago on Earth, when life was just beginning? There would surely not have been as many types, because much of the diversity of viruses today relates to specific hosts, and before the evolution of the first living cells there were no hosts—or were there? What if viruses preceded the evolution of cells? We will return to this thought below.

So if there are a lot of them out there, how many different ones are recognized today? Virologists have long been interested only in those that cause disease, but this has been changing recently, and many new types are discovered each day. Today the International Committee on Taxonomy of Viruses (ICTV), the ruling body of those studying the diversity of viruses, recognizes more than fifteen hundred "species" of viruses, but this number is ludicrously low. Virologists have identified another thirty thousand "strains" of distinct viruses, and this number must be a huge underestimate of what is out there. The reality is that we have a very poor idea of the diversity of Bacteria, Archaea, and Eukarya, and we know that each of these domains is riddled with parasitic viruses—perhaps millions of different kinds, in fact.

What is clear is that the great difference in the biology of a virus and a cell has made a mockery of utilizing normal taxonomic practice on viruses. A species is usually defined as a group of individuals that are reproductively viable. But viruses do not reproduce in the same way that cells do, let alone have sex of any kind. Defining a virus as a specific kind of species is probably a great disservice to the concept of species. Viral taxonomy, in fact, has been characterized as an opinionated use of data. Nevertheless, taxonomy of viruses is moving along at an accelerating rate as the result of the large amount of sequence information coming from many labs looking at the nucleic acids in various virus taxa. (Unfortunately, these are not the same sequencing practices used to differentiate Terroan life—ribosomal RNA.) The Linnaean taxonomic structure has now been applied to them, but with a few interesting additions.

The primary characters used to classify viruses are actually few in number. They include the type and organization of the viral genome. Is it RNA or DNA, and what kind of each? Second, how does the virus in question replicate? Finally, what are the morphology and chemistry of its body wall or coating like? Viral species are discriminated on the relatedness of the genome (based on sequencing), the natural host, the type of cell within the host that it normally invades, and its danger to the host, known as its pathogenicity; its mode of transmission; and the chemical properties of its virion, among other things. Once we start organizing viruses in this way, we start seeing a huge variety of forms.

Another way of looking at viruses is through their hosts. For our purposes, we can ask where on the tree of life viruses project and if specific viruses are found on only specific parts of the tree. This latter statement seems ludicrous at first. Viruses are parasites, and all cellular parasites are host-specific. We might thus expect that specific virus groups would be found on very specific branches of the tree of life. A viral type that infects mammals, for instance, would not be expected to infect a bacterium. But is this true?

We can best answer this question by first looking at the large-scale divisions of viruses, using the criteria listed earlier. The biggest distinction between viral kinds is those with DNA and those with RNA; this breaks up the viral world into two major hemispheres. After that, however, there are many other ways of subdividing the diversity of known viruses, and rather than burden this narrative with a long list of viral types, we can perhaps best do this job by summarizing the viral types and their characteristics in a table (see Table 2, p. 54).

Another unusual aspect of this new view of viruses is how they evolve. There seems to be a schizophrenic aspect to being a virus. Individual viruses specialized for parasitism of one or a few kinds of cells must constantly evolve to keep up with the defenses of the host cell. Earlier we quoted the statement that the swarms of viruses around and in each cell of life on our planet causes a huge selective pressure on the living cells. Natural selection will thus cause the cell being attacked to try new methods of defense against the viruses if they are too destructive. This type of coevolution between predator and prey, or parasite and prey, is well known. But the viruses seem to show another kind of evolution, one distinct from this day-to-day battle with the host. The virus carries two types of structural and functional components: one necessary for the day to day, and the second, far more conservative aspect of "self": the way the genome is packaged, things that go back to a long-ago time when viruses first evolved. Bamford thinks that these observations indicate that viruses and cellular life are intimately and anciently linked. He has proposed that viruses form lineages that extend from the root of the tree of life to all branches of the tree. The implication is that viruses were there when cellular life formed. This leads to a further interesting question: Were viruses opportunistic forms that early on took advantage of the earliest cellular life, riding through time as parasites, or might they have had an even more intimate acquaintance with the Terroans—in fact, as an agent in the formation of life as we know it? There is a final aspect about viruses made explicit by Bamford. At the conclusion of his article he states: "If the above reasoning is considered it follows that there is a separation between the viral and cellular world. . . . Perhaps we should consider formally dividing life into viral and cellular, where the cellular one is formed of the current domains and the viral world of its lineages." Later in the same paragraph Bamford states: "The idea of virus lineages would also mean the viruses were present before the separation of the domains of life at the very root of the tree of cellular life." Amen to that.

I encountered this sentence after I had already decided to put viruses on the tree of life and was much heartened to find a similar point of view from one who knows viruses professionally and intimately, the result of a life of study. There are enormous implications from all this about what life is, how we classify it, how we understand its evolution. There are clearly two vastly different kinds of life on Earth: viral life and cellular life. From this we must again change the tree of life that we have been building so far. And from this we can propose a new hypothesis for how life evolved.

Was the origin of virus life mono- or polyphyletic? Is there a LUVA, a last universal viral ancestor, analogous to the LUCA of cellular life?

The evolution of viruses

There are a variety of viruses on the planet today. While it has long been assumed that viruses have a single origin, there is an emerging consensus that they are polyphyletic, that they came from independent sources. This is the view I favor. Some viruses might be degenerate RNA organisms, and thus are very ancient, while others might be a more recent parasitic form.

Viruses can store their genetic information in six different types of nucleic acid which are named on the basis of how that nucleic acid eventually becomes transcribed to the viral mRNA capable of binding to host cell ribosomes and being translated into viral proteins.

In the table (+) and (−) represent complementary strands of nucleic acid. Copying of a (+) strand by complementary base pair forms a (−) strand. Only a (+) viral mRNA strand can be translated into viral protein. These six forms of viral nucleic acid are:

Type of Genome	Description	Group of Viruses
(+/−) Double-stranded DNA	The (−) DNA strand is directly transcribed into viral mRNA.	Most bacteriophages, papovaviruses, adeno-viruses, herpesviruses
(+) DNA or (−) DNA	Once inside the host cell, it's converted into dsDNA, and the (−) DNA strand is transcribed into viral mRNA.	Phage M13, parvoviruses
(+/−) Double-stranded RNA	The (+) of the (+/−) RNA functions as viral mRNA.	Reoviruses
(−) RNA	The (−) RNA is copied into a (+) RNA that functions as viral mRNA.	Orthomyxoviruses paramyxoviruses, rhabdoviruses
(+) RNA	A (+) RNA is copied into (−) RNA that is trans-cribed into viral mRNA.	Picornaviruses, togaviruses, coronaviruses
(+) RNA	The (+) RNA is reverse transcribed into (−) DNA that makes a comple-mentary copy to become (+/−) DNA. The (−) DNA is transcribed into viral mRNA.	Retroviruses

Table 2. *The Variety of Viruses and Their Specific Kinds of Genetic Code*

If all these many kinds of viruses are alive, where do they fit on our tree? They are certainly not like us Terroans as I have defined us. But they are related to us. Some kinds of viruses may actually be "living fossils" coming down through time from a period before life on Earth had DNA. These are RNA viruses. As long as twenty years ago RNA viruses were considered living fossils from that time.

So living viruses present us with a classification problem. We need to define a new taxon. Like our Terroans, it must be placed at a level higher than that currently in use. We need to define a second dominion.

So let us get that over with. Just as earlier, we can formally define this new dominion here and then get back to our cooking.

Dominion Ribosa, **Ward, 2005**
Definition: *Life composed of carbon chemistry with water solvent with RNA as its genome.*
Included groups: **Domain Ribovira, Ward, 2005**
Definition of Domain: Encapsulated life with RNA as genome. Includes RNA viruses.

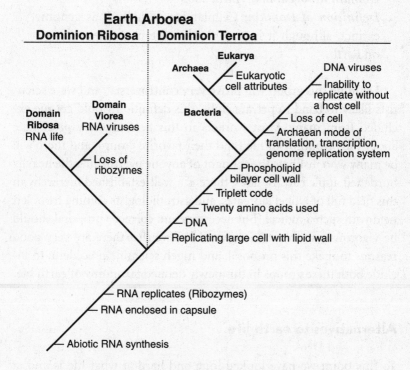

Earth Arborea
Dominion Ribosa ⋮ **Dominion Terroa**

- Eukarya
- **Archaea**
  - Eukaryotic cell attributes
- DNA viruses
  - Inability to replicate without a host cell
- **Bacteria**
  - Loss of cell
  - Archaean mode of translation, transcription, genome replication system
- **Domain Ribosa** RNA life
- **Domain Viorea** RNA viruses
  - Loss of ribozymes
  - Phospholipid bilayer cell wall
  - Triplett code
  - Twenty amino acids used
  - DNA
  - Replicating large cell with lipid wall
- RNA replicates (Ribozymes)
- RNA enclosed in capsule
- Abiotic RNA synthesis

A new hypothesis for the evolution and classification of earth life.

It is not only viruses that should be considered as living. In chapter 4 I describe the numerous reasons that suggest that some sort of RNA life preceded our familiar Terroans. An organism using RNA as its genome would not fall onto the Terroan tree. While all attention has been focused on how this perhaps mythical beast brought about the eventual evolution of us Terroans, no one has yet classified it among the current categories of earth life for the simple reason that it cannot be included in any traditional category. Suddenly we are confronted with the need to define a new group of life, one that cannot be placed within any category of earth life. So here we come to the another conclusion of this book (so far): A second domain in Ribosa of life must be identified, if RNA life existed.

> ***Domain Ribogenoma,*** **Ward, 2005**
> ***Definition of Domain:*** Cellular life with RNA as genome, extinct, although it is possible that extant species still exist on Earth.

Defining this group will prove very controversial, and were scientists likely to band in private clubs, this definition would get me excluded from many. Most workers in this field are very passionate about the taxonomy of life (and their favorite group), and there will be many who just hate the thought of anyone overturning the heavily burdened apple cart. Besides, there is a well-established hierarchy in this field full of Nobel laureates, and astrobiologists coming from left field with such a radical (but necessary and correct!) proposal should be vigorously snubbed. However, I believe that there are very good reasons to make this proposal, and much scientific precedent, to include both these groups in this newly defined taxonomy of earth life.

Alternatives to earth life

To this point we have looked long and hard at what life is and at what earth life is. As we have seen, life on Earth has been very successful in colonizing much of planet's surface, even down a kilometer or two into its crust, as well as far up into the atmosphere. The range of conditions goes from hot to cold. But even this great range of temperature, pressure, and amount of oxygen, acidity, and other factors influencing life are not as extreme as we find elsewhere in the solar system. Only a tiny fraction of earth life might be able to exist on many planets and moons of our solar system. But what of non-earth life? In the next chapter we'll look at what life might be in terms of chemical diversity.

Chapter 3

Life as We Do Not Know It

So I tell my students: learn your biochemistry here and you will be able to pass examinations on Arcturus.

—George Wald

We have looked at what known earth life is. I have also tried to make the case that viruses are alive and must therefore be placed on the tree of life, necessitating a radical revision of the tree. In a way, viruses might be considered aliens, because they are so different from the life as we know it that is cellular life. But let us now look at other possible life chemistry that is clearly alien compared with our own. Two factors seem imperative in this discussion: First, any such alien life should be chemically permissible; second, it has to be able to form or be formed. In this chapter we look at what life might be that is not earth life, the life as we do not know it of this book's title. In chapter 4 we'll examine the second part of this: how life might be able to form.

How can we systematically approach this subject? One way is to begin with aliens that differ from our newly named Terroans of planet Earth in the smallest details and then range farther afield toward more unfamiliar life-forms, things that would do justice to the wildest fringes of the UFO/alien-in-the-freezer contingent. For example, we could break these aliens down into the following categories: DNA with language or syntax change, or proteins with a different assemblage of amino acids; DNA with chirality reversal; life without DNA as its genome (such as RNA life, protein life, or possibly a protein genome life). Then things get a little more diverse: carbon life using some other solvent (such as ammonia), life based on some other element than carbon, and the variously proposed "exotic" types of life. But is this the only way, or the most logical way, to categorize life as we do not know it?

To do justice to a handbook of alien biology, which this chapter aspires to be, we need a bit more detail. A handy way of covering the necessary bases can be gleaned in any biology book. Perusing the table of contents from the text used at my university in its freshman biology course, one can quickly see what alien life would need, based on what our own familiar life would need. But first, we should ask, is there any science in studying something that has not ever been observed?

Obstacles to the study of aliens

A small group of outstanding scientists are now seriously thinking about what they call weird life. As I state in the Preface to this book, this new direction of research is backed by the most exclusive club in the world, the U.S. National Academy of Science, by the National Research Council, and within NASA by a visionary administrator of astrobiology, Dr. Michael Meyer. The main instigator among the scientists is John Baross, of my university, who is part of both the Carnegie and University of Washington nodes of the NASA Astrobiology Institute. Baross has thought deeply about this subject, and the following list of problems associated with studying alien life comes largely from a presentation that he made in June 2004 and that I recapitulate here with his permission.

Baross points out that the biggest obstacle to understanding what life could be is that we do not yet understand the origin of life on Earth. Until we do, it will be hard to understand accurately how life might form under different conditions from those attending life's formation on Earth—if it did form here at all instead of arriving via airmail from some other site or origin. Perhaps the biggest problem stemming from origin uncertainties relates to biochemistry. Until we can understand the full range of the biochemistry of carbon-based life, we will not be able to arrive at a vigorous and scientific study of the possibilities of life.

Our search for life is based on a "follow the water" methodology. We see this in the Martian missions undertaken by both NASA and the European Space Agency. Because carbon and water are key to life on Earth, we search for both in space. In essence, we are searching for parallel habitats. But Baross makes an important point: How parallel do they have to be to support life? Parallel habitats can currently be defined as those with water, a source of carbon, key nutrients (such as compounds with readily accessible sulfur, nitrogen, and phosphorus), and physical and chemical limits that resemble those that can be survived by extremophilic microbes on earth today—a range of temperature, pressure, acidity or alkalinity, and radiation that are not inimical to earth life. Because we have so little idea about what the limits of nonearth life might be, since we have no good description of nonearth life, we are in a place where there may be too many unknowns about earth life to be able to study and understand alternate carbon-based biochemistry. It has always seemed to me to be the ultimate Ph.D. question for the occasional biochemistry exams that I have to referee at my university: "So, candidate, in ten minutes or less sketch out for me an alternate carbon-based life-form that is chemically permissible but not based on DNA or RNA." The successful respondent should be given first a Ph.D. and then a Nobel Prize for Chemistry, it is that important and difficult a question.

Where might the parallel habitats exist in our solar system? Later I will visit the list that Baross favors: the subsurface of Mars and Europa; the dry surface environments of Mars; the salty ice cover of Europa; the deep, oligotrophic depths of Europa; and the hydrothermal systems that may (or may not) exist, or have existed, on Mars and Europa. In September 2004, NASA made its strongest pronouncement to date that early in its history Mars had bodies of water that were the size of the Great Lakes or of small seas on Earth—at least. If there were a hydrothermal system in this early Martian sea or lake, we would have an environment parallel to that of Earth, old and new alike. Last on Baross's list is a place that is becoming a favorite site for possible life of almost all those who seek aliens, Titan.

The two most important ingredients for earth life appear to be carbon and water. So the obvious starting point deals with life without carbon or life without water, or both.

Carbon/water chauvinism—how universal is CHON life?

In chapter 1 defined life in an abstract sense. Let me get more practical and approach life as a mechanic would (or as a chemist would, if that chemist were involved in the monumental effort to synthesize organic life). In this way we can then approach the possible differences that alien life might have and still stay alive. So, what is needed?

There are two major components: solid matter, which makes up the structure of life, and liquid. The first part, the need for solid structure, is self-evident. But why liquid? Why not purely solid life? That liquid is present is evident. We humans are about 70 percent by weight water, as are most animals. Plants use less fluid in their bodies, but not much less for most plants. The water in our bodies is used for several highly important functions. Most important, perhaps, it acts as a solvent, which, as we will see, is necessary to maintain and allow many of the crucial chemical reactions that keep us alive. But water does more than that: It acts to regulate temperature, to provide a substrate for nutrients to float in within the cell, and to provide chemical equilibrium within the cell. To keep our water in liquid form, we must maintain body temperatures between 0° and 100°F under normal pressure conditions (some microbes can have high internal temperatures and maintain their water as liquid if under high hydrostatic pressure caused by ocean depth).

Earth life, our familiar Terroans, is what is called a carbon-based life-form, a form that can be called CHON life, for its preponderance of carbon, hydrogen, oxygen, and nitrogen. This may seem a curious definition since there are lots of other elements involved in the familiar earth life: phosphorus, iron, and sulfur, to name but a few. But carbon makes up the backbone of most structural components of our bodies. Using a different structural element is a staple of science fiction and none more so than so-called silicon life. Why isn't earth life made of silicon, instead of carbon, and could there be a theoretical silicon life-form?

A look at the periodic table of elements suggests that silicon should work. After all, it is right underneath carbon, one row down, and thus should share many of the chemical properties of carbon. Like carbon, it is very abundant in the solar system (and galaxy, for that matter). Because of its location, it can, just like carbon, combine with four hydrogen atoms. But here the similarities end. The most important differences between silicon and carbon (those that would affect the kind of life using these elements) lie in the very nature of each element's chemistry. The most relevant aspect is the strength of chemical bonds formed by carbon compared with those formed by silicon. Since chemical reactions are all about bonding and breaking bonds, bond strength is hugely important. For example, bonds between silicon atoms can be broken much more easily than carbon-carbon bonds, so any structure composed of a long series of carbon in a chain will be much stronger than a chain of silicon. On the other hand, silicon-hydrogen and silicon-oxygen bonds are stronger than those in compounds using carbon instead of silicon, and this makes it less easy to produce rings or chains of silicon atoms, something that life needs. Carbon chemistry is dominated by chains, rings, and branched chains, and these units of structure figure prominently in both the structural and metabolic needs of carbon life. Silica fails at these chemical levels.

Stars made all the carbon that exists in the universe. Our carbon is no exception. One of the reasons that this star stuff is so good at making life (and lots of other carbon compounds) is that it bonds well to other atoms and other molecules. Any carbon atom can form four bonds, and the bonds it makes with other carbon atoms, or with oxygen or hydrogen atoms, can be very strong. Here is an example of the strength of these bonds: At the temperature of liquid water, a carbon bond can endure for thousands of years. No other element forms so many bonds that are so strong. It is akin to the old Tinkertoy sets (which, not coincidentally, are sometimes used to portray molecules; the best builders were the round disks of wood with four holes in them, just like a carbon atom).

But does that mean that some sort of silica life is impossible? Not in the least, it turns out. Silica certainly forms stable compounds, as witness the large assemblage of stable rocks that are classified as silicates (such as granite). This is due to silica's love of oxygen. When present, silica bonds with oxygen from highly stable compounds— but silicates react with only a small number of other chemical species and thus seem unsuitable for the myriad chemical needs of life—at least as we know it and in environments like those on Earth. Silica life of any sort would not work on Earth. But theorists have concluded that under conditions of very low and very high temperatures, some kinds of life with silicon within it might be viable. There is a lot of very cold and very hot real estate in space beyond the Earth, so we had better give this type of life a good look.

The most imaginative workers thinking about alien life are Gerald Feinberg and Robert Shapiro, who, in their 1980 book *Life Beyond Earth*, thought so far outside the box that they often lost sight of the box. They have weighed in on silicon life and suggest that it could exist at very high temperatures, where the crystals that make up such sheet silicates as mica dissolve and turn into liquid state. The two also suggest that a form of evolution could act on liquid silicates and potentially create a primitive form of life. Such life might live deep in the earth (the mantle, for instance). What is unclear is how such life would metabolize and replicate—also requirements for life, as we have seen.

Liquid and life

We know of one chemistry that works, and anyone suggesting that carbon-based chemistry is the only way to make life is lambasted, in spite of the fact that there has never been a working alternative presented in any detail (to be fair, there is still no mathematical or even chemical expression of our kind of life that takes into account all its complexity and workings). So will carbon life be common in the universe? Many thinkers on the subject of life, N. R. Pace being just one example, suggest that we should expect to find carbon-based organisms on virtually every planet that our starship *Copernicus* lands on in the nearest million light-years. Given CHON life, we should *expect* to find biological components made up of peptides, sugars, and nucleic acids—just like our own life. This is what chemist William Bains calls a parts list of life with "builders' specifications." Bains suggests the parts list of Terroan life is one narrow subset of the huge hardware store that life can shop in to make itself. Bains also makes a sensible suggestion. Just as NASA does in searching for life, look for the liquid or, in this case, as Bains suggests, look *at* the liquid, the importance of which he describes: "[I]t is the nature of the liquid system in which chemistry arises and not the limitations of chemistry itself, that will direct the biochemistry of non-terrestrial living things." Let's examine his argument.

Bains postulates that liquid environments where biochemistry might take place are common in the universe. We have lots of them in the solar system alone. The water or water/ammonia oceans of Callisto, Europa, and Titan; the ethane/methane lakes potentially on the surface of Titan; the subsurface water of Mars; even small droplets in the Venusian atmosphere might qualify. If we want to get really alien, there may be liquid nitrogen geysers on Triton, a large moon of Neptune. But Bains points out that our type of liquid, fresh to moderately saline liquid water, is but one kind of liquid where chemists would allow life to be, at least according to their rules. If we accept this, we are pretty quickly in the land of alien life, since the chemistry of a liquid in which life lives will be the major controlling factor in its chemistry.

It turns out that there should be a variety of liquids in our solar system and, by inference, in other stellar systems as well. Oceans will be either surface or subsurface and vary in chemistry on the basis of their distance from the sun. This is caused both by the way our solar system originated (most water and volatiles are found beyond the orbit of Mars) and by the relative temperatures to be found as a function of the distance from the sun. In the figure below, these relationships are shown:

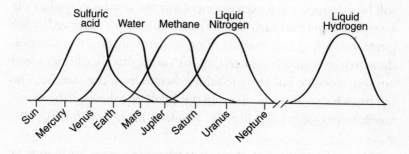

Liquid solvents that might be found as a function of distance from the sun. This same relationship should hold for other planetary systems besides ours. Thus we might expect characteristic kinds of life, dictated by solvent properties, as a function of distance from the star. Bains, 2004

As we move farther from our planet Earth into space, the nature of any liquid becomes dictated by the originating chemistry and by the extreme cold that we find as we move away from the sun. The question then becomes: Can Earth's biochemical systems function in these more exotic liquids? If not, what might a chemistry of life be like? There is no doubt that the earth's biochemical systems are enormously adaptable, but there are limits—especially with regard to the solvent chemistry and temperature. Many Terran biochemical molecules will not be viable in the low-temperature liquid such as water/ammonia solutions at very low temperatures, while the high pH found in many of these oceans (such as that of Europa) would quickly break DNA and RNA apart through the process called hydrolysis.

Varieties of CHON life

What makes up a plausible alien? Not some whacked-out science fiction extravaganza, not even an alien animal, but something simple: an alien microbe. In this section let us suggest some plausible aliens by going through the various components of our familiar DNA life and then in an orderly fashion deriving something unfamiliar. At first this will be a disappointing exercise, especially for anyone steeped in the lore of sci-fi and the nearly infinite possibilities of life suggested by the genre. But I find that even limiting our imagination to microbes, the search is ultimately exhilarating, for a sense that alien life is indeed not only possible but also probable emerges from the exercise. The sad truth is that all of the aliens in this book, and there will be many, are microbes, or microbelike. No Wookies, Predators, Vulcans, etc., etc., etc.

To begin with, we need to look at what life is if we are to look at what it might be. To recap, our familiar Terroan variety of CHON life uses DNA for its genome, builds its proteins from twenty specific amino acids, has several basic metabolic systems for energy extraction, and uses a bilayer cell membrane made up of the fatlike substance known as lipids. So, let us see what might be changed.

Changing genetic code

Perhaps the simplest way to make an alien would be to change DNA slightly. Our familiar DNA is a double helix made up of two long strands of sugar, with the steps of this twisted ladder made up of four different bases. The code is based on triplet sequences, with each triplet either an order to go fetch a specific amino acid or a punctuation mark, like "stop here." Within this elaborate system there are many specific changes that could be made—at least theoretically—that would be "alien" yet might still work.

Perhaps the simplest way to change DNA would be to change its code or alphabet. There is no a priori reason that the current base triplet has to be specific for the amino acid it now codes for. Another simple change is to add more letters into the alphabet, and several groups have done this. Steve Benner did this a decade ago by making a code that had twelve bases per letter. The four traditional bases of the DNA ladder—adenine, thymine, cytosine, and guanine (ATCG)—were still used. But instead of a sequence of three, or triplet, being specific for each amino acid, a sequence of twelve was used.

Another way to change the language or coding is to use new or additional bases within the DNA molecule itself. Recall that our kind of life uses only four particular bases (the rungs on the DNA ladder). Steve Benner's team succeeded in producing a DNA-like molecule containing *six* different kinds of steps instead of the traditional four. That is indeed an alien DNA. I wonder if a creature entirely made out of this kind of DNA would be alien in some unexpected fashion. Benner's new DNA look-alike was able to go through five generations of replication. These experiments clearly show that DNA could come in many varieties of languages, and changing the base coding or base number would be simple. It would be interesting to know if early in earth history many separate kinds of DNA with different codes competed against each other. Is a twelve-nucleotide DNA code more or less efficient than our familiar three-nucleotide coding? Was there a square-off among a whole series of DNAs, or was the first to achieve this grade of organization the winner, suppressing a variety of equally or even more effective competitors through some sort of incumbency advantage? These questions will be looked at in the future. But clearly there is a whole suite of aliens that could be produced through these changes alone.

Is this new DNA dangerous to normal earth life, by the way? After the announcement of his finding Benner stated: "We doubt that our artificial DNA would survive for an instant outside of the laboratory on our planet. But a six-letter DNA might support life on other planets, where life started with six letters and is familiar with them. Or even DNA that contains up to 12 letters, which we have shown to be possible."

Changing the code for proteins is simple. But it turns out the ATCG bases that DNA uses for its code are not the only chemical bases that can be used. A group led by Floyd Romesberg has succeeded in using altogether different bases. In this DNA, the sugar "backbones," the sides of the helical staircase, remain the same (deoxyribose sugar), but the steps are changed. Instead of ATCG, up to twenty different bases were successfully substituted. The unexpected finding was that not only could this new code system call for the usual twenty amino acids, but entirely new amino acids could also be coded for. Romesberg is well on his way to producing a DNA molecule that is made up of "unnatural" base pairs that will function in a living organism. This will, by anyone's definition, be an alien.

But is life with such a minor change really an alien? In a wonderful coincidence in which real life meets TV, an *X-Files* episode featured a character using a six-nucleotide DNA, just like that produced in the lab by Benner and his group. The character was, by the way, an extraterrestrial by *X-Files* standards, at least, and what better authority to pontificate about what is or is not alien? Benner has thus produced the most important part of an alien, an alien genetic code.

Changing the backbone of RNA

Changing the code of DNA is one way to make an alien. Another is to perturb another part of the molecule, in this case the sugar backbone of DNA and RNA. Both use a sugar backbone, deoxyribose for DNA and ribose for RNA. How about a different sugar? Several scientific groups have synthesized artificial RNA that uses a different sugar from ribose. The most successful was a sugar called hexose. Soon it was found that a number of distinctly different RNA-like molecules could be synthesized. No one yet knows if these new molecules can carry out the processes that RNA can: act as messenger, enzyme, and code. What would a creature using another sugar in its DNA be like?

Changing or adding proteins

Earlier we documented the finding by the Romesberg group that its new unnatural bases could code for amino acids that are not normally found in earth life. This is not the only group that has experimented with making proteins alien to that used by common earth life, us Terroans. In 2001 two different groups persuaded a bacterium (the common *E. coli* found in our guts) to work with a new suite of amino acids. The results, by groups headed by Lei Wang and Volker Doring, got the bacterium to use a twenty-first amino acid. It was a different twenty-first for each group, suggesting that non-Terroan life might use a suite of amino acids not used by us. In each case the use of this twenty-first amino acid marks this newly tinkered bacterium as an alien by my definitions of what life is.

Changing chirality

All Terroan life uses proteins with a single chirality, a term used to describe whether an asymmetrical organic molecule is right-handed or left-handed. A simple change from left-handed to right-handed would certainly make an alien.

Changing solvents

We saw in an earlier section that there are many kinds of liquid on planets and moons. One logical way to make an alien would be to change the solvent, the liquid required for life. But even here we are facing a great unknown. Would DNA work with a solvent other than water? Consider the ramifications of alternatives to water.

We must judge the effectiveness of a solvent first and foremost by its ability to do what a solvent does: dissolve things. This occurs because of the chemical activity of a polar molecule. Water, with a chemical formula of H_2O, looks like a pair of Mickey Mouse ears with a large oxygen atom on one end attached to the earlike and smaller hydrogen atoms. With the end of the oxygen dominating one end of the molecule, and the hydrogen ears the other, water has distinctly different chemical properties from end to end.

Water has many other properties that aid in the maintenance of earth life or any similar forms of life. It is a very good solvent, it is an appropriate medium for conducting chemical reactions, and the substances that it dissolves separate in positive and negative ions, which can aid ongoing or further chemical reactions. Water has a great ability to store heat; it acts as a reservoir for heat dumping. Bodies of water thus tend to stabilize their surroundings against rapid temperature swings, in the same fashion that a coastal region is buffered against rapid temperature swings that are experienced in a desert; the nearby ocean or lake stabilizes the climate through the heat-trapping properties of water.

Water is not the only solvent, of course. Other solvents that might work in a biological system include ammonia, methyl alcohol, hydrogen sulfide, hydrogen fluoride, hydrocyanic acid, and hydrogen chloride. Each might be found in some biological being, and thus each might conceivably form a whole class of life. But water is much better than any of these as a solvent, and because of this, given the chance, life might normally proceed with water over any other solvent.

The ability to break apart chemicals, however, is not the only property that a good biological solvent needs to have, and it is for these other behaviors that we might see life resorting to something other than water. For instance, water works better than anything else over the range of temperatures at which it remains a liquid—from 32° to 212°F. But only a small fraction of the universe enjoys such a temperature range. Most of the universe is either much hotter or much colder, and in these places other solvents would be superior to water. A good biological solvent should remain liquid over a large range of temperatures; if it freezes or boils, any life it rests in will almost always die. (We shall get back to freezing, where this may not always be the case.) Some of these (such as ammonia) have very low freezing points and thus might work in very cold places. Others, such as concentrated sulfuric acid, have very high boiling points (higher than water) and might be useful for life trying to exist at very high temperatures. In sum, a solvent for life must be able to dissolve enough substances to allow life to have a chance either to form or to continue, and it must be made up of simple components that can be found as well as remain in the liquid phase in the environment in which the given life form is forced (or chooses) to live.

Substituting proteins for nucleic acids as an information storehouse

Life on Earth uses proteins for structure and enzymes and nucleic acids for storing information. Both are large and complex molecules. Could there be life if one used substitutes that were simpler, made up of structures composed of simpler molecules? Of these two, proteins are the simpler, and if they could conceivably replace nucleic acids for the information-storing part of life, perhaps a life-form could be envisioned. Many proteins in extant earth life are also extremely complex, and we have to ask if simpler proteins could have the catalytic properties of some of these very large protein molecules. Our life is complex and efficient. Could there be life less complex and far less efficient but life nevertheless? A very fertile field in the study of alien life (at least of theoretically possible aliens) is in looking for simpler rather than more complex kinds of organisms compared with earth life.

Life in solid, liquid, or gas?

Here on Earth we humans (and probably the majority of earth life) live in air. While there are millions of marine species, it now seems clear that terrestrial diversity, especially among insects, is greater than the overall diversity of marine species. But our ancestry is made from water, and we air dwellers (or land dwellers, if you prefer) are but water-filled bags moving around. This leads us to the question, Could there be life in gas or in solid in the same way that we are water creatures?

Let's look at solid first. Life, as we know it and don't know it, must be chemical in nature and must be able to undergo chemical reactions. For this to happen, the various molecules capable of chemical activity have to be able to move around one another yet also be in close proximity. This is where any theoretical solid life and gas life are at a great disadvantage. In a solid, there is close proximity but little movement, while a gas has lots of movement but no proximity. For these reasons we can eliminate life in a solid or gas from serious consideration.

Specific and plausible CHON aliens

Here I'll try to put together a list of possible alien life-forms, rather than simply list structures or characters that could be changed to produce an alien.

RNA life

This might not be an alien at all in the sense that it lives (or lived) off the earth. As we have seen in chapter 2, all Terroans might have RNA life of some sort as an ancestor. What would it look like, and where might it live?

Ammonia life (also called ammono life)

The concept of ammonia life goes back to 1954, when the great chemist and origin of life specialist John Haldane conceived of an alternative biochemistry in which water is replaced by ammonia. Ammonia, like water, dissolves many compounds and remains in a liquid phase over a very wide range of temperatures, a range that even increases under pressure. Ammonia has a much lower boiling point than water, a property that makes it an interesting candidate for a solvent of low-temperature life or at least life that can live at a lower temperature than Terroan life. Also, there is no shortage of ammonia in our solar system. There would be major differences between ammonia life and Terroan life, of course. Such life would have to have a very different kind of outer cell wall or membrane, since liposomes (a fatty component that is a major structural part of Terroan cell walls) dissolve in ammonia. And metabolism would be very different by necessity as well. Ammonia cannot be used in Terroan life. The hope of making an earth bacterium somehow using ammonia and water as an internal solvent (if only to demonstrate that such life could exist) is doomed to failure, even though no less a personage than Carl Sagan once considered trying to concoct such a life-form in the laboratory.

Could there be a viable ammono life-form? Steve Benner and his colleagues think that such life is theoretically and chemically possible. Where Terroan life exploits compounds using carbon-oxygen bonds in metabolic pathways (specifically the unit known as carbonyl), a workable metabolism using carbon-nitrogen bonds seems possible, according to Benner. So here is an alien that seems plausible and worth studying or perhaps synthesizing.

Acid life

In chapter 8 we look at the possibility of life high in the clouds of Venus. The problem there (for Terroans, that is) is the very high acidity. While some microbes on Earth today live in acidity equal to or even greater than that which would be encountered in the high-cloud environment on Venus, another solution would be an alien biochemistry that deals better with the nature of the Venusian cloud layers. Living among these aerosols, or small droplets of acid, might be some type of life.

Non-CHON life

So if not CHON life, what? Let's now return to the silane or silicon life we briefly flirted with and give it a longer look.

Silane (or Silicon) life

Elsewhere we have extolled the virtues of carbon and dismissed the possibility of silicon life. Part of the disdain for potential silicon life comes from a backlash against its science fiction roots; it is as much a staple of the genre as barroom brawls and pistol duels are to westerns. There are many reasons for this, and on planets like Earth, the evolution of silicon life is probably impossible. Even if it could get started, it would surely be soon outcompeted by us carbon-based life-forms. But earthlike planets may not be a dime a dozen; in fact, to quote two dissidents named Peter Ward and Don Brownlee, they may be rare. But just because a good earth might be hard to find does not mean that there is no other real estate that might be suitable for life. One of the perplexities to me is that the astrobiological fraternity has concentrated on thermophile extremophiles so much, and while there are indeed hot spots in the solar system and the universe at large, there is a lot more cold than hot out there. It is in the realm of the cold that we need to explore biochemistry to find the outer fringes of what life might be. Eventually finding a CHON alien of some sort, if it ever happens, will be an amazing and transcending experience for humanity. But finding something even more exotic would be even more interesting. Let us look at the exotic and possible—silicon beasts in the cold.

Cold is a challenge for biochemistry—CHON biochemistry, that is. Reactions run slower, and compounds that need to be dissolved in a solvent have the nasty habit of not dissolving or doing so at such slow rates as to be essentially useless. By the time we hit $70°K$ ($-200°F$), virtually nothing dissolves in any solvent. Where might we find such cold temperatures, places so cold that nitrogen becomes a liquid rather than a gas? Triton, the moon of Neptune, for one. In liquid nitrogen (and in liquid methane or liquid ethane sitting around on the surface of Saturn's Titan, for instance), organic molecules, such as methane, acetylene, and carbon dioxide, will dissolve, but larger organic molecules will not—with one notable exception. Enter the silicon life-forms, championed in separate articles by William Bains and Steve Benner and his colleagues.

Silicon can form analogs to our carbon-based alcohol. These are called silanols, and they are soluble in a wide variety of solutes, at a huge range of temperatures, including the very cold temperatures at which nitrogen remains liquid. When they dissolve, they populate the solute—be it ethane, methane, or liquid nitrogen—with analogs to the carbon-based organic molecules. They are large molecules, and once dissolved, they can form polymers, the chained-together molecules necessary for life. And since they are silica rather than carbon-based, they have weaker bonds to break. They thus have greater reactivity than carbon-based compounds. This is a problem at earthly temperatures, making silica a liability as a molecule central to life, but in very cold environments, this would be a very good thing indeed for any life-in-waiting. Because of this, less catalytic efficiency is needed to perform necessary chemical reactions in the very cold by this theoretical silica life.

For these two reasons, the ultracold environments of our solar system would be a more favorable place for some type of silicon life than for carbon life.

Life, as we have seen, is far more than simple chemical reactions. What of the structural components of silicon life? As any good geologist knows, there are numerous silicate minerals that form very stable (too stable) chemical compounds. But silicate rocks are composed of large chains or sheets of silicon bonded to oxygen. Silicon can also form stable polymers, a building block of life, by Si-Si molecules, called silanes. Silicon can also bond with carbon. It is thus a mistake to consider any possible silicon life as being composed mainly of silicon, just as it is a mistake to think of carbon life as mainly carbon. In both cases numerous other atoms can tag along, and even complex carbon compounds can bond to the silica chains as side branches, thus allowing silanes to form diverse and complex molecules with side chains making them analogous to carbohydrates, nucleic acids, and proteins, all the stuff of carbon-based life. Silicon compounds can even form ring structures like the carbon compound benzene and like sugars, thus illustrating silicon's versatility.

What about metabolism as it relates to our theoretical silicate creature? How will silicon chemistry be modified to yield energy? Where many of the metabolic systems of carbon life shuttle protons in energy-harvesting systems (like the ATP-ADP system), silicon might do the same not with protons but with electrons. There could be light-activated electronic effects, mimicking photosynthesis, as well as other energy analogies.

Where might we find silicon life? Not in water and not in water-ammonia solutions since these solvents would quickly destroy the complex si-organic molecules. But as mentioned earlier, not all bodies in the solar system have water and/or ammonia. On Titan we found ethane/methane lakes, and on Triton, perhaps an ocean of liquid nitrogen. In such places we might find silicon-based life. Would we even recognize it as such? The very cold might make it pretty slow life compared with the speed of chemical reactions on our warm Earth. Would we even recognize life moving at very slow rates?

The last aspect to consider is whether there is a plausible pathway by which a silicon life-form could originate. Once again, on the basis of the nature of its chemistry, this has been shown to be chemically permissible.

Finally, how far might such an organism evolve? Could we expect multicellularity? Would there be an evolution toward greater complexity? Could such an organism ever be anything other than a microbe? Let's follow these unanswered questions with an even more intriguing one. It is clear that earth life has vastly changed the environment of the earth's surface. The evolution of oxygen, of multicellularity, of plant life and trees with roots: The list is enormous. So important has been the contribution of life to the nature of our planet that a hypothesis, the Gaia hypothesis, has been formulated. Its most extreme adherents believe that the connection between life and the planet show that the planet itself is alive. We will look in more detail at this idea in the section "Implausible aliens" that follows; it pretty much sums up my views on the subject. But there can be no doubt that life changes its environment. That said, could we even envision how silicon life might change its environment? If there were silicon life in the methane lakes on Titan or in the liquid nitrogen of Neptune's Triton, how would the long-term existence of such microbes have affected their respective worlds? What is the equivalent of the oxygenation that our world has gone through or the nearly disastrous snowball-earth episodes that life triggered here? And if Titan and Triton have similar life-forms, could one live on the other? We chauvinistic carbon units think that panspermia, the idea that life can go from place to place, deals with us. Could it be that it is the silicon life-forms that fly from place to place on comets and asteroids, populating the universe with its most dominant form of life, silicon life? My colleague Bob Pappalardo thinks not; his assessment is that little might escape Triton's gravity well, reducing the chance of panspermia between these two bodies. But there is still a chance.

Silicon/carbon clay life

Another radical variety of potential alien life is that envisioned by the geologist Alexander Cairns-Smith, one of the most creative astrobiologists around. He has imagined an entirely new kind of life, crystal life. His view that silicon-rich clay might be alive is both reviled by life scientists and considered a brilliant insight. It is as follows: "For a picture of first life do not think about cells, think instead about a kind of mud and assemblage of clays actively crystallizing from solution." He suggests this in the context of a hypothesis about how CHON life evolved on earth through a silicon life precursor. Here I put this interesting idea in the context of plausible alien life.

The life-form Cairns-Smith proposes so audaciously is nothing less than a growing crystal. He envisions small (when I say small, I mean small; we are not talking about some science fiction rock creature, but a life-form based on a flake of clay invisible to the naked eye) crystals of clay that actively grow and evolve as they do so. Cairns-Smith singles out the clay known as kaolinite as a prime example of how this type of life might exist.

A kaolinite crystal grows by accretion (a layer-by-layer application of new minerals) on its faces, one of which is broad and flat compared with the others. A typical crystal of kaolinite has three layers of oxygen atoms stacked like three flat layers of oranges in a box. The oxygen atoms, oriented in this way, are held together by smaller atoms lying in the crevices between the oxygen (the oranges) and bonded to them; these interstices are made up of a plane of silicon and a plane of aluminum atoms, so in all we have five planes of atoms making up our crystal. To complete our clay particle, there will be some small hydrogen atoms tacked on to the oxygen. This is obviously a sophisticated structure, with both a top and a bottom side. The whole shebang can be thought of as a carpet (this is Cairns-Smith's analogy). The growth of the crystal is upward, with a carpet put on top of the first, another on top of that, and so on, until our now-visible clay particle is composed of thousands of the three-molecule-thick carpets stacked one atop another. If this growth proceeded in perfect fashion, each carpet with all its molecules in perfect place, there could be no opportunity for life in this system. But while growth normally produces flat sheets stacked one upon another, there are often irregularities, caused by the addition of rare elements and by copying errors. Other deviations from uniformity come about by the process of twinning, a geological process found in most crystals, causing them to split and branch.

Not only does our stack of carpets grow upward, but the whole carpet can become two side-by-side carpets. The crystal is thus replicating in the instances in which the carpet splits into two upward-growing pinnacles by splitting into two growing stacks, with the stacks now side by side. Our "organism" has now replicated, one of the criteria of a living system. But it takes more than this to be alive. Where does the evolution come in? Where are the genes? Cairns-Smith has answers to these too.

The evolutionary scenario envisioned by Cairns-Smith has numerous stacklike crystals of clay all in the same environment and all competing for "food"—the atoms of silicon, oxygen, and hydrogen dissolved in water that surround the crystals. When these atoms come out of solution to increase the size of the solid crystal, we say that growth has occurred. Cairns-Smith hypothesizes that some stacks, because of their shape, gathered the atoms needed for growth more efficiently than did others and grew faster. There was thus competition for resources. Some particular shapes are then selected for, and a crude sort of evolution takes place, with the ions' coming out of solution being the metabolism. While at first this description sounds like anything but life, this point of view may be our carbon/water chauvinism showing.

Pretty soon there are a variety of crystal "life-forms" out there, growing away. Where are their "genes"? The top layer of the growing crystal is the gene; it holds all the information necessary for the "organism" to grow, just as a DNA molecule holds the information necessary for an organism to grow.

There is far more than simple shape in determining winning and losing. The crystals are selecting for many details, including the presence of microchannels and distinctive microshapes to aid growth, as well as chemical composition. Now we see a world of growing clay crystals. But would we really see them? Again, this is taking place at the microscopic level, and the microbial world is invisible to us. The next step in the evolution of the clay life is that some of the crystals start adding organic molecules to their lattices. Again, if the addition of carbon compounds aids growth rate and survival, they will be selected for. Soon we have clay crystals with organics on them, and perhaps some of these figure out photosynthesis and other processes that carbon compounds do well. Cairns–Smith sees an organic takeover, whereby the channels and pumps of the clay minerals are replaced entirely by the organic compounds, the clay genes by nucleic acids. Eventually the takeover is complete, and all forms with crystal parts go extinct—or so the story goes for the earth.

If this scenario is plausible—and some very good scientists seem to think that it is—why would there always be a takeover? Are there environments where the clay creatures are favored without, or with only partial, organic components? And how would we recognize this kind of life if we fell over it? Its rate of metabolism would be much slower than ours in many environments.

There are several variants on the clay life idea. One is based on the fact that clay minerals have repeating units and sites that could hold other molecules, including organic molecules. John Bernal, a crystallographer, suggests that the flat sheets that characterize most clay minerals could serve as a template that could help assemble the organic molecules that then combine to form organic life. These would be CHON life, presumably, with an assist from minerals.

So I offer this vision as one plausible alien. Finally, could such aliens be on earth today? How would they be tested for? How fast is a "life" in clay? Could the rates of change be so slow that we could not perceive them as living? This is a prime example of a kind of alien life that we would probably never recognize as such or discover.

Implausible aliens

Let us jump from the plausible to the less than plausible. Plasma life is one of these. A staple of the *Star Trek* genre is life formed not of solid, gaseous, or liquid matter but of what is sometimes called the fourth state of matter, plasma. Plasma is not solid, liquid, or gas. Seems about as reasonable as most of the long-shot (but imaginative!) ideas on *Star Trek*—or is it? In 2003, physicist Mircea Sanduloviciu and colleagues in Romania produced small spheres of plasma that just might show lifelike characteristics. They introduced a spark into plasma of the gas argon and created small spheres with negative charge on the outside of the spheres and a positively charged inner side. Some of the spheres took in more argon and enlarged, while others split. But what about evolution? Here the spheres failed. Not life.

Living planets and the Gaia hypothesis

Let us look at another really implausible idea—that a planet itself is a form of life—and scrutinize the components that make up a greatly influential hypothesis that the earth may be alive (and if one planet is, surely there are others). In the defense of this idea, it is clear that planets have life spans, and some would say life cycles of a sort: They form, evolve over time, and ultimately are destroyed. It may even be said that they metabolize, after a fashion. But can it be said that they are alive? This is the premise of the Gaia hypothesis. A living planet would certainly qualify as life as we do not know it, and such a being would tax the tree of life methodology to the max.

People throughout time have championed the concept that the earth is in some way alive. Most recently, this view has taken on new credibility because of adherence to it by a collection of world-class scientists, led by the British scientist James Lovelock. In a series of books published in the 1970s and 1980s, Lovelock focused this view. This hypothesis is based (mainly) on the idea that the biomass of life self-regulates the conditions on the planet to make its physical environment (in particular, the temperature and chemistry of the atmosphere) more hospitable to the species that constitute its "life." The most extreme form of Gaia theory is that the entire earth is a single unified organism; in this view, the earth's biosphere is consciously manipulating the climate in order to make conditions more conducive to life.

The Gaia movement has spawned an interesting new discipline called Earth System Science. Yet the analogy of a living human and a living planet is easily misused, as evidenced by the extreme nonsense promulgated by many self-described Gaians. Our planet is not alive. Planet Earth is a closed system with respect to material—essentially we do not receive new material from outer space but continuously recycle what is already present—but an open system with respect to energy, whereas all organisms are "open" systems with respect to both. Humans and almost all other organisms do not last very long without a constant intake of new material.

The question of convergence

Powerful insights in science can usually be recognized with very little data. One such is the theory of convergence in evolution. It is immediately recognizable that fish, dolphins, and ichthyosaurs have similar body shapes and tails and that birds, bats, and pterosaurs all have similarly shaped wing structures. Anyone can understand that all these structures evolved within relatively unrelated groups of organisms to perform similar functions under similar conditions. This is a hallmark of evolution. It is also relevant to our discussion about what life as we do not know it might be like. The extreme end of this line of reasoning would have you believe that there is no life *other* than earthlike life because all life will converge on the same DNA type of life that we have. In a huge and detailed book, *Life's Solution,* Simon Conway Morris of Cambridge University has looked in painstaking detail at the process of convergence and its relationship to potential alien life. His take on the ubiquity of convergence, not only among the earth's species but as well for any life beyond Earth, is very relevant to our discussion about how alien aliens might be.

Morris's argument goes back to his near-death struggle with Stephen Jay Gould about evolution. While the sociology of this major quarrel among two of the giants of late-twentieth-century biology and paleontology is in itself fascinating, the main argument bears discussion in our context. Gould used the powerful metaphor that he called replaying the tape: that if the history of the earth and the solar system were somehow replayed again, we would see a different assemblage of animals and planets. Morris, however, argued that while many details of evolution would change, the overall look and feel of animals and plants would remain familiar because of the process of convergent evolution. On a planet where gravity, atmospheric composition and pressure, and seawater salinity are the same, goes the argument, convergence would produce a suite of creatures that, while possibly having different ancestry, would probably look familiar. There are not too many ways that biology can produce wings or fins and not so many ways to fly or swim most efficiently. Yet this argument between these two men was waged at the level of our familiar vertebrate cousins, rather than at the level of microbes. Thus Morris's take on the way that organisms can harvest light energy is fascinating.

The light from stars must be one of the most ubiquitous and

surely utilized energy sources in the universe. Because light energy can be caught and used to maintain the order that is life, one might expect there to be a large number of ways to utilize this resource. Life as we know it uses a molecule that has been named chlorophyll. We can ask, Is the chlorophyll molecule used by our Terroan life one way or the *only* way to harness and harvest light for life? The chlorophyll molecule must have been an ancient invention because we find it in bacteria that appear very early on the tree of life, as deduced by RNA comparisons among microbes. One group using chlorophyll, the cyanobacteria (often called blue-green algae, even though they are not algae), may be very ancient indeed, and some workers, such as Joe Kirschvink of Cal Tech, suggest that they are true living fossils, coming down to us from at least 3.5 billion years ago. Cyanobacteria and higher plants all use a very similar process to tap light.

The chemical processes used to harvest light using the chlorophyll molecules are, like life itself, very complex. There are a series of steps and necessary enzymes, and one of these, with the unwieldy name of D-ribose-1, 5-biphosphate carboxylase (happily shortened to RuBisCo, or rubisco), is especially important. Curiously it does not work well in the presence of oxygen. Chlorophyll shows other drawbacks, including its narrowness in utilizing the spectra of light that are out there. But for all its drawbacks, no one can design what would be an alternative to it. It seems that carbon-based life, even of another code, would converge on the chlorophyll molecule if it wished to harvest light. This view is strengthened by the discovery that the chlorophyll molecule appears to have been evolved by widely divergent stocks of earth life. Convergent evolution has led to this solution, and this may be the case even for DNA. It may be ubiquitous among life in the universe simply for being the best thing available. If you give life enough time for natural selection to weed out the less efficient varieties, it may be that the earth model of chlorophyll, and perhaps DNA as well, is always selected.

Chlorophyll is not the only example of convergence, of course. Let us think about locomotion—especially among microbes. The ability to move is of fundamental importance to the survival of many microbes. To avoid toxic habitats or find resources, microbes show a surprising ability for locomotion but a very low diversity of methods for movement. While there are some wonderful adaptations, like the use of small crystals of magnetite that allow all magnetic bacteria to move toward magnetic fields, concerted movement among most other microbes is largely propelled by flagella, long whiplike structures, and cilia. Again, even in divergent stocks, the internal construction of these structures remains remarkably homogeneous. This suggests that convergence has been at work and is a powerful and perhaps universal property of evolution—among aliens as well as earth life.

The acceptance of the power of convergent evolution is not limited to morphological structures. As we have seen, metabolism is a key component of life, and we can expect that there are self-similar energy sources throughout the universe. Light, chemical energy from reduced carbon, the use of hydrogen in certain rocky habitats all might drive convergence toward similar enzymatic pathways, which in turn will drive life toward fundamental similarities. Those who argue for a huge diversity of alien life ignore convergence—at their peril, it seems, if we are reading the history of life correctly. Simon Conway Morris has, through his remarkable treatise on its ubiquity in earth life evolution, shone a light on what aliens off the earth might be like. They might very well look much like us, even while having radically different internal chemistry. Perhaps silicon life, if it exists, looks much like carbon life.

A summary of aliens

So where do we stand in this zoo? Let us take stock with a table of aliens.

Name	Scaffold Element	Gene Material	Solvent	Scaffold Element Source	Energy Source	Possible Terrestrial Habitat	Solar System Habitat	Plausibility
Terroan	Carbon	DNA	Water	CO_2, other organisms	Many	Most	Mars, Europa, Titan	It exists
RNA life	Carbon	RNA	Water	Organic molecules	?	?	Titan, Mars, Europa, Earth?	High. Once existed and and may still exist on Earth
Protein life	Carbon	Proteins	Water	Organic molecules	?	?	Titan, Mars, Europa, Earth?	
Ammonia life	Carbon	Nucleic acids or proteins	Ammonium	CO_2? Other?	Sunlight	None		
Acid life	Carbon	Nucleic acids	Water or ammonium				Venus clouds, Jupiter clouds?	
Silicon life— silanes	Silicon		Ethane	Crystal fluids			Titan, Triton	
Silicon/clay	Silicon		Water				Ancient Earth? Mars?	

Table 3. A Summary of Aliens

Detecting aliens

How will we know an alien if we find one? One of the truly disappointing aspects of the NASA planetary program has been the absence of life detection equipment on so many of the space probes sent to other planets or moons in our solar system. In NASA's defense, such equipment is bulky and difficult to build. Furthermore, there are very few signs of life that are unequivocal. Still, the Martian rovers have no life detection capability, and neither did the *Huygens* probe that landed on Titan.

So what would an adequate instrument package be that could both detect life's presence on a planet or moon and determine whether it was our kind of life or not? The omnipresent (at least in this book) Steve Benner has some suggestions.

Benner has pointed out that a genetic molecule such as RNA or DNA must have a repeating charge on each unit of its backbone. Where every phosphate-based group attaches to the molecule there will be a charge, and the phosphates have the important property of keeping the DNA or RNA molecule from folding in on itself and thereby becoming nonfunctional. Repeating charges are not something that one bumps into among the dead, and Benner argues that instruments could be built to detect this in water. Will the same effect be found in other solvents? Probably. The beauty of this system is that the instruments need look only for repeating positive or negative charges. Even for molecules of life very different from our DNA, this same characteristic will apply.

Finally, we know that there are biosignatures of our kind of life. Specific biochemical molecules have been found in old rocks on Earth that could have been made only by life and therefore leave a chemical fossil or biomarker of the long-ago life. These organic molecules change and degrade through time, heat, and pressure but do so in predictable ways. These pathways are recognizable and detectable.

Being able to exist does not necessarily allow existence

In this chapter we have looked at many kinds of potential alien life. But being *able* to exist does not mean that something *will* exist. It is becoming clear from the long study of how life began on our own Earth that the conditions necessary to make life (which includes the presence of necessary building blocks) might be far more difficult to produce on a planet than the conditions necessary for life on that planet, once (and if) it evolves. We can imagine many cases in which the crucial components are never brought together on an otherwise livable planet. This caveat surely goes for alien life as well as earth life. Thus now let's look at mechanisms discovered by astrobiologists and others that might bring about the existence of life.

Chapter 4

A Recipe Book of Life

The origin of life is a science writer's dream. It abounds with exotic scientists and exotic theories, which are never entirely abandoned or accepted, but merely go in and out of fashion.

—John Horgan, *The End of Science*

My wife is a sensational cook. Part of that excellence might come from the fact that she follows recipes to the letter. No cheating at all, with slavish devotion to using the ingredients called for, in the quantities called for, in the sequence called for. And no skimping on the ingredients! Fresh, fresh, fresh! She can even bake. Of course, another important trick is finding recipes that work and finding recipes that work on the particular equipment she has to *work* with—our stove, oven, particular frypans, all the outrageously expensive paraphernalia particular to our kitchen. When I cook, on the other hand, I make it up as I go along and never follow recipes—with pretty predictable and disastrous consequences. Thus I have developed a profound respect for all those who have perfected the various arts of cooking and documentation of recipes. As you might suspect, this aside on cooking has a purpose here. There is an obvious analogy between the formation of life and the cooking of a complicated dish. The really complicated ones involve many individual smaller jobs ("cook so-and-so, and then set it aside"), so that the ultimate "cooking" involves the assembly of many individual precursors, some involving prior baking, frying, etc., with the whole shebang finally being assembled through some complex chemical process of heating or even heating and pressure. The whole thing is so foreign to me that I can't even find the right words. Sometimes I think it simpler just to consider all this cooking as some miracle taking place out of my sight. Much easier that way. But the chaos in the kitchen afterward, the mess left behind, puts the lie to that little fable. Some pretty complex chemistry has been going on, and leaving evidence of itself for cleaner-uppers to deal with. There are no miracles in cooking.

The formation of life might well be like this: The many components (membranes, nucleic acids, proteins) are first cooked up according to a variety of simple to complex recipes, and then all these ingredients are assembled and cooked in some very specific way. This recipe might call for thousands of ingredients assembled over millions of years! Would it be so simple as: Assemble a bunch of organic molecules in just the right proportions, bake at 45°C for two million years, and *voilà*! Life! The task in this chapter, now that we have made a list of potential *kinds* of life, is to look at potential *pathways* to life. Documenting all the potential pathways to life, if we assume that there are more than one (sadly we do not even know if that is true), is a large task, and to organize it, I am going to write my own cookbook in the pages of this chapter, starting with recipes for some of the components of life as we know it and life as we don't and then finishing with several plausible recipes for making Terroans, all the while noting that in spite of much work by many cook-detectives, we still do not know the precise recipe for making a Terroan. As cooking involves specific energy inputs in most cases (you cannot fry a cake or make Kentucky Fried Chicken by baking), I will also describe some of the various "ovens" where life on Earth, at least, may have been produced. So, with no further ado, here is the *Planetary Guide to Recipes of Life.*

Going to the store, and the store coming to the planets

Luckily for life, most planets and moons have at least some of the ingredients to make many varieties of the life mentioned in the previous chapter. In chapter 2 I discussed the way in which planets form, and we saw that early in solar system history a steady rain of comets plagued most bodies in the solar system. On the basis of many observations, planetary geologists have concluded that comets contain many organic compounds, including some amino acids. Thus, in addition to the already present elements necessary to make life, there has been a delivery system in effect throughout not only our solar system but probably in all planetary systems, as a part of the planetary formation process.

So, most of the ingredients to make life were present on our own planet some four billion years ago and probably on the other planets and moons that might still harbor life in the solar system: Venus, Mars, Europa, and Titan. Yet some of the most crucial ingredients, at least for Terroan life, were not readily present and had to be separately cooked up. And they were crucial! Try making bread without yeast. That is analogous to making Terroan life without the complicated workhorse molecule RNA. So let us begin our cookbook by looking at the various suggested recipes for making RNA.

The problem of cooking up RNA

Amazingly, one of the major criticisms of RNA life, one of the alternate life-forms introduced in the last chapter and the hypothesized last common ancestor of all DNA life, is that it probably did not exist *because it would have been impossible to build RNA through natural chemical processes.* Paul Davies notes: "As far as biochemists can see, it is a long and difficult road to produce efficient RNA replicators from scratch. . . . The conclusion has to be that without a trained organic chemist on hand to supervise, nature would be struggling to make RNA from a dilute soup under any plausible prebiotic condition." Once RNA has been synthesized, the path toward life is open because RNA can eventually produce DNA. But how the first RNA came into existence—under what conditions and in what environments—became the central problem facing chemists. As Nobel laureate Christian de Duve notes in his book *Vital Dust,* there just seemed no way to see how RNA could come about naturally: "We must now face the chemical problems raised by the abiotic synthesis of an RNA molecule. These problems are far from trivial."

The very difficulty of making RNA has led several scientists working on this problem to suggest that life may have begun with some alternative to RNA, which was then later taken over by RNA when a lot of the "heavy lifting" in life's synthesis from inorganic to organic was over. The more delicate RNA and DNA molecules were later substituted in. The problem is temperature; in elevated temperature, RNA and DNA quickly fall apart. So much for all that dinosaur DNA in amber touted in the *Jurassic Park* movies. You cannot make amber unless you cook it in the oven of geological time, and any enclosed DNA is cooked to something else in short order.

It is not just elevated temperatures that create problems for synthesizing life from inorganic components. I have already extolled the virtues of water as a solvent, virtually canonizing this substance as something that most life probably needs if it is going to be life. But for the creation of RNA and DNA life, water seems no more friendly to most of the organic components making up earth life than it did to the Wicked Witch of the West in the *Wizard of Oz;* for both, it dissolves things in very nasty fashion. While water is a necessary ingredient for many amino acids, the very characteristic so useful in maintaining existing life, its ability to dissolve compounds readily, works against it. The bases making up the ladder of DNA and the steps of RNA are chemically transformed by water, and many polypeptides (long carbon chains necessary for building the framework of cells) will dissolve into their constituent amino acids. Water is no less fierce for those all-star molecules DNA and RNA; parts of both molecules will break apart in water, requiring a constant input of energy to stop this destruction. Is there a way around this? The problem seemed intractable until even a few years ago, but luckily a new recipe using borax soap (of all things) has become available.

The borax recipe for making RNA

Years ago, before he was president, Ronald Reagan was the host of a television oater titled *Death Valley Days*. An industrial-strength soap that used borate minerals for its cleansing action, 20 Mule Team Borax, sponsored the show. The young Reagan valiantly sold truckloads of this soap to a national television audience; my parents bought it, and it was disgusting stuff. Who knew that all that borax soap seems to have been a key ingredient in the natural synthesis of RNA and thus life itself? It turns out that minerals in the borax family may be a pathway to a key ingredient of earth life, RNA. So let's look at a perhaps prime recipe for making life.

As mentioned earlier, the hardest step of all was making RNA. It was a real head scratcher. All the rest of the steps from a chemical or prebiotic soup to a cell could be understood, and from RNA to the DNA life of today (our lovely Terroans) was also understood and possible to do in the lab. But going from sugars and bases to RNA has defied laboratory work. Moreover, it was not just the inability to replicate this step in the lab. Some big guns, heavyweight Nobel Prize and National Academy types, came out showing just how tough a job this process was in nature. It turns out that RNA is a very wimpy molecule, large, complicated, and very easily destroyed. DNA, in comparison, is much sturdier. The great old man of synthesis himself, Stanley Miller, for years at Scripps, a bastion of research into the origin of life, flatly stated in an article in 1995 that the instability of ribose sugar, the key ingredient of RNA, under a variety of conditions made it unlikely that RNA supported the emergence of life. In his view, then, RNA could have come only later, *after* life had already started. Miller was especially critical of those suggesting that RNA could have formed, for instance, in the heat and violence of the hydrothermal vent environments. The sad truth is that water attacks and breaks up nucleic acid polymers (strings of smaller molecules), making it difficult to envision RNA forming out of some prebiotic soup that was water-based. The major trouble is that there are many steps in making RNA, and each step would require different conditions or a different chemical environment. Our poor forming RNA molecule would have to be express-mailed from place to place on the earth, going from one site to another. Not too likely, said the gurus. And it is not only the problem of getting to RNA, but even how to arrive at one of its major components, the sugar ribose, which is the R of RNA. The brilliant biochemist Antonio Lazcano has described this problem thus: "The RNA-world model confronts several serious challenges, including the lack of a plausible primitive abiotic mechanism to account for the formation and accumulation of ribose." If even making the sugar of the far more complex molecule we call RNA is so difficult, what is the chance that nature would stumble onto the hugely complex molecule itself? So, such pessimism must mean the problem of RNA's original formation is intractable. Happily, new work suggests that is not the case.

The cook who perfected this dish or at least the most promising solution to how RNA could form on a planet or moon is Steve Benner, of the University of Florida, whom we have met already in these pages, and who is hereby awarded the title Master Chef of RNA.

Benner came at this problem from a chemist's point of view. But he is not your ordinary chemist. He knows his minerals, and his key insight was that certain mineral groups surely present on the early earth in environments where RNA could have first formed paved the way for ribose formation, the key ingredient of RNA. Ribose is very unstable and reverts to a brown tar unless it is kept cold. Tar is of no use in the formation of life, but ribose is. What happens to sugars in general, and the sugar ribose in particular, is all too apparent to anyone (like me) who has tried to bake a cake but left it in the oven too long. Eventually the cake becomes asphalt, to use a metaphor from Benner himself. Sugar turns into something quite unpleasant under heat. So a major problem is that ribose and electrical sparks, presumably the energy source of the early earth, are not compatible. While an almighty lightning storm energized Frankenstein's monster in the eponymously named movie, it would not help in this case. But clearly RNA *did* originate in some natural fashion, unless it fell from space to Earth already formed, but that just passes the problem back in time and elsewhere in space.

Thinking about the problem, Benner concluded that some aspect of the earth's early environment must have protected the forming nucleic acids in such a way as to shield them from the heat and other deal breakers present in most environments. His breakthrough came from the realization that ribose would chemically bond to certain minerals in such a way that even heat would not turn them to tar. There are several such minerals in the large class of minerals called borates. Benner then tried some experiments, using a specific borate mineral named colemanite, which is found in Death Valley. Pretty ironic too, considering that St. Ronald Reagan, no evolutionist, hawked the stuff that might have been the key ingredient in the evolution of the first life itself, for Benner found that in the presence of borate, simpler organic molecules that are common both on Earth and in space (on comets, for example) combine to form complex sugars, including ribose. Here was a way to make the stuff.

Soon Benner found that other borates, minerals such as ulexite and kernite, also function to preserve the structure of ribose under heat conditions that, without the borates, would turn the ribose to tar. A doorway of understanding was opened. It is no longer unthinkable to imagine the formation of nucleic acids in high-temperature environments. This work was published in early 2004 and is still in its infancy. Benner and his group may have solved the single biggest hurdle to elucidating the recipe for Terroan life.

So, say that we have made some RNA and have set it aside for later, grander cooking. Now what recipe might we use to get to life itself, if we assume that a naked strand of any old RNA is not alive, a debatable concept in some circles? Let's move on to a more grandiose recipe.

The recipe for garbage bag life

With our RNA cooked and now safely sitting on the counter awaiting our need for it, we can start thinking about moving from this vital ingredient to a final dish, cellular life. Here another cooking analogy is apt. Anyone dealing with the prospect of roasting a large turkey will find that cooking the bird in a cooking bag speeds up the process. One idea about the pathway to life, irrespective of the environment in which it is taking place, is to make life in the equivalent of a cooking bag. Freeman Dyson, whose work we looked at in chapter 1, has come up with an ingenious recipe for making Terroan life. Dyson has worried about how the two main functions of life, replication and metabolism, could come about in the same nascent organism: "There are accordingly two logical possibilities for life origins. Either life began only once, with the functions of replication and metabolism already present in rudimentary form and linked together from the beginning, or life began twice, with two separate kinds of creatures, one kind capable of metabolism without exact replication, and the other kind capable of replication without metabolism. If life began twice, the first beginning must have been with molecules resembling proteins and the second beginning with molecules resembling nucleic acids." From this startling image, Dyson goes on to describe a world where there were nucleic acid creatures (essentially RNA life), necessarily obligate parasites preying on the second kind of life, the protein creatures, and pirating the products of protein metabolism for their own replication. Such a definition seems very close to what viruses do today, only the nucleic acid creatures would have been much cruder, without the sophisticated action of modern-day viruses. Dyson calls this idea the double origin hypothesis. He uses another, far more descriptive analogy: garbage bag world, where proteins come first and nucleic acids second. In cooking terms, garbage bags (primitive but empty membrane spheres) are put in some kind of soup cooker (a warm little pond? a hydrothermal vent?) with RNA and other replication ingredients added, and out comes cellular life. Would that it were so simple.

The sequence of events leading to life where cells (Dyson's garbage bag; I prefer to call it a cooking bag) come first, followed by proteins within the cells, culminated by the formation of nucleic acids to direct it all, was first theorized by the Russian biologist Alexander I. Oparin in his 1924 book *The Origin of Life* (obviously a popular title, as this is the third book with that specific or nearly that title we have already referred to). Dyson has expanded on the Oparin theory and in his book *Origins of Life* (note the emphasis on more than a single origin in the title, since he champions his double origin hypothesis) describes the sequence of events in colorful, if gritty, style: "Life began with little bags, the precursor of cells, enclosing small volumes of dirty water containing miscellaneous garbage. A random collection of molecules in a bag may occasionally contain catalysts that cause synthesis of other molecules that act as catalysts to synthesis other molecules and so on. Very rarely a collection of molecules may arise that contain enough catalysts to reproduce the whole population as time goes on. The reproduction does not need to be precise. . . . The population of molecules in the bag is reproducing itself without any exact replication." Occasionally, according to Dyson, the bag splits into two bags. He concludes: "As soon as garbage-bag world begins with crudely reproducing protocells, natural selection will operate to improve the quality of the catalysts and the accuracy of the reproduction." Paraphrased, garbage bag world posits small, spherical membranes filled with molecules, none as complicated as RNA or DNA. These chemicals begin catalyzing reactions with one another, some of which eventually cause the cell-like spheres to divide. Eventually, a genetic code using nucleic acids comes about within the cells. So here we see cells first, enzymes second, and genes last. And this is a recipe that should work not only on Earth but beyond Earth as well. Following this recipe, we can envision that both earth life and some varieties of alien life could be fashioned in this way.

This, then, is Dyson's vision of the long-ago earth. Somewhere, perhaps amid the hydrothermal vents or maybe in warm little ponds or somewhere else, microscopic bags of proteins, their lipid membrane walls separating them from the surrounding heated seawater or pore fluid of some sort, interacted with a different group of aliens, perhaps nearly naked strands of RNA. Perhaps there were ribosomes already formed. In any event, the two types of assemblage merged into a single cell.

The RNA life recipe

Our recipe for cooking bag life is only one entry in the making life cookbook. Here is a second. In this recipe, genes come first, in the form of RNA, but the difference from the garbage bag model is that the naked RNA is itself living rather than the inert ingredient in the garbage bag recipe. Only later is the RNA enclosed in a bag of some sort, and eventually DNA takes over from RNA as the genetic code-bearing molecule. This is the so-called RNA world first theorized by Manfred Eigen and his colleagues in 1981. It has genes first, enzymes second, and cells third. This is a different recipe from that posited by Oparin-Haldane-Dyson, in which the first component to be assembled was a functional cell. These are really different, radically different recipes of life.

That some sort of early earth life could have existed using RNA alone seemed impossible for many years, for all current life uses enzymes to catalyze the reactions that both make new cell material by constructing proteins and undertake the crucial chemical reactions that yield energy, the important function we have named metabolism. But without some DNA molecule giving directions, how could enzymes be constructed? And without enzymes, the reasoning went, there could not be protein synthesis. So it was long concluded that naked RNA could not be alive. But this belief was overturned by the discovery that RNA not only can act as a messenger but can *perform chemical catalysis as well.* Thus the possibility that primitive cells may have carried out all the functions of replication and metabolism with RNA alone was given a strong boost by the Nobel Prize–winning work of Thomas Cech and Sidney Altman, who discovered that RNA can indeed act as the enzyme necessary for catalytic activity. In this situation we have RNA acting as the storehouse of a genetic code as well as a catalyst—two entirely different jobs. These RNA enzymes were named ribozymes and led to the concept of the RNA world. Paul Davies, in *The Fifth Miracle,* supposes that some sort of "soup" of RNA molecules could evolve by a Darwinian type of mechanism of evolution into the first living cell. The RNA world was a time, according to Freeman Dyson, "when RNA *life* was evolving without the help of protein enzymes."

I have added the italics to this last sentence. While the sentence is of great importance in the understanding of how life evolved, it is the definition of "RNA life" that I find so striking and so important. In a book dealing overtly with the definition of life, Dyson has identified a kind of life that cannot be included in the new category called Terroa that I defined earlier. Davies makes a similar assertion, suggesting that whatever these RNA accumulations were, they certainly showed characteristics of life and indeed (if we assume that they existed in some form) should be considered as having been alive. This is justification for my proposal in chapter 2 of classifying RNA life as an extinct dominion of earth life (see page 56).

Our cookbook is getting thicker, but we are yet not done. Here is another recipe for life. It might end up as our kind of life, but then again it might be a recipe for alien life.

The clay life recipe

Here is another pathway for the evolution of earth life (Terroans, at least, but I imagine a whole spectrum of life could be brought into being with this recipe), coming from a geologist, of all people. We can call the finished product clay life. (In chapter 3 we discussed clay life as a possible type of alien life-form, as it might be. But it may have been an intermediate step leading with its own recipe to the finished product of earth life, and hence it is included here in the recipe section as well as in chapter 3.) At about the same time that Eigen and colleagues were formally proposing their RNA world hypothesis, Alexander Cairns-Smith made the ingenious suggestion that naturally occurring minerals contained in various clays might have served as a primitive genetic material before nucleic acids were first evolved. The irregular distribution of aluminum and magnesium ions would be the carrier of genetic information. These embedded ions on flat crystal surfaces of clay would form areas of varying electrostatic potential that could both absorb molecules and catalyze chemical reactions. The crystal could thus produce a RNA-like function of guiding the formation of amino acids and proteins and perhaps aiding a metabolism of some sort. Somewhere along this progression, organic chemistry would take over from clay chemistry. In the old legend, glancing at Medusa turned life to stone. Perhaps we should call this pathway the reversed Medusa path to life.

This order of things has clay first, enzymes second, cells third, and genes last of all. We can envision late in this process a small protocell containing a microscopic flake of clay, somewhat analogous to the silica chip of a computer, except that the clay is not only the software but also the hardware construction site. The change occurs when RNA is found to be a better genetic material than clay and, through natural selection, takes over.

The biggest problem with Claymation as an origin of life is that once again it has, as its Achilles' heel, the formation of RNA. Once again we have a system whose most difficult aspect is getting to RNA.

The recipe for pyrite life

Ditto above. Pyrite life also provides part of a recipe leading to earth life and potential alien life. It is another kind of rock life that has been proposed, life living in obligate fashion on crystal surfaces, in this case pyrite rather than the clay of Cairns-Smith. Like clay life, pyrite life, if it existed, would require its own dominion of life.

We thus have several recipes for life. Sadly, we do not yet know which of these recipes was used for earth life. In fact, garbage bag life and RNA life might be mutually exclusive. But that leads to some fascinating speculations. Did one or both of these pathways to life get used on earth? If so, will each result in a particular and different kind of CHON life?

A new recipe: viral stew

Here is a recipe that I have made up. (With my mea culpa about cooking prowess, one would think that I would stay out of the kitchen. Oh, no.) Anyway, this recipe might make Terroan life from a stew. At the end of this cooking experience, if all goes well, we should have a batch of living, thriving Terroan cells that we can fish out, and we'll then throw the rest of the stew away. But it also makes something that the other recipes have ignored: viruses. It's not my fault that they are such a problem; who can keep a leash on one's kids anyway?

We have already prepared some parts of our stew, and there are many kinds of different and individual ingredients floating around. The "stock" of our stew is water, but there are so many kinds of organic compounds and molecules in it that it is more like an organic muck. Among the ingredients there is some newly produced RNA as well as a huge diversity of protocells of at least two very different-sized classes. The smaller group of protocells (again, they are probably nearly all different other than sharing a size) is composed of small bags of polypeptides (primitive proteins), and they are indeed very small—virus size, in fact. There are lots and lots of different kinds of these, but they all share the characteristic of their cell walls being protein. The other protocell group is hundreds to thousands of times larger, and like the smaller group, it is made up of many different kinds, each differing by the chemical compositions of its outer walls. These are made up of many kinds of lipid membranes, some single-layered, some bi-layered, some with a protein or two stuck in them—a whole zoo of them, in fact. So we have a host of big bags, a host of little bags, and zillions of spaghetti-like strands of naked RNA that cannot replicate themselves. They form inorganically and then fall apart. But the larger of the protocells, those made of lipid, can replicate, in a fashion. They are tiny bubbles, really, and sometimes split into tinier bubbles. To this point this recipe is very much like the garbage bag recipe championed by Freeman Dyson. But now we bring in some new ingredients.

There is a form of competition among all three of these components and a form of natural selection for greater efficiency. The next step in this process is a merging of the RNA strands and the smaller of the two classes of protocells, the ones made up of proteins. These are so small that they snugly fit around the RNA strands, causing the RNA to fold onto itself. Within this enclosure, for the first time, the RNA can replicate. It is very small and very crude in its replication, but when it replicates like this, it is, in fact, forming primitive ribozymes, the organelles used by Terroans to make proteins. Many mistakes are made, and the mistakes become the stuff of natural selection. But at this stage we can call this RNA/protein bag alive. The protein becomes the capsid, because with this advance is born the first virus, which is also the first life on Earth, an RNA virus that differs from all of today's viruses in being able to replicate its internal RNA by using a very primitive ribozyme system. It is simultaneously RNA life and a virus, but not very efficient as either or at being alive. But it is life by any definition. The protein wall can allow the ingress of solutes but no larger macromolecules. Our virus usually starves to death, or "dies" from a lack of energy. No food, no energy, but once in a while a successful replication. Over and over among the soup these things come together and fall apart. Meanwhile the larger of the two protocells, that with the lipid cell wall, is much better at evolving energy-generating systems with a potential difference over the membrane barrier. Occasionally one of these bags can take in large particles and harvest energy. Among those taken in are the tiny protein-coated RNAs. It is a great place to be in this protocell.

So now we have our virus in the large garbage bag, a bag filled with organic goodies. There is much more "food" here than outside because the organics are concentrated in the garbage bag, compared with the outside world. Our little virus can take some of these organics into itself; tiny bits of ribose and amino acids are especially appreciated. Replication of the virus happens more often and more successfully. Once in a while the protocell splits in two, carrying some of its now-captive viruses with it. Inside this protocell our viruses continue to evolve through mutations. They can make a few crude proteins that add to the survival of the larger protocell— probably most important, enzymes involved with transport and energy harvesting. Cellular life is eventually born. But our original virus, occasionally still captured in a crude fashion, does not die out. It evolves. Its protein coat changes in composition to allow the virus ready access into the interior of the ever-larger cells, those with increasing efficiency of protein formation. The virus no longer needs to replicate itself within its own tiny capsule. It loses its ribozyme system. Why bother? Like a blind cave fish, it no longer needs "organs" that once were crucial for survival. Our viruses (for they are coevolving with the many variety of cells also proliferating) no longer need to replicate using their own biosynthesizing machinery. They coerce the wonderful enzyme-rich cells to do it for them.

Up to this point all are part of the RNA world. Both the protein-coated forms and the new lipid wall forms use RNA genomes. But the evolution of DNA changes everything. It arises within the evolving cellular life, and with it a comes a second kind of virus, using the same route: There arise inside the large cells capsid-enclosed DNA strands, rogues with instructions on making more of themselves, gaining ever more efficiency. And here is a further point that, while controversial, seems to be borne out by looking at how genomes of modern cells incorporate segments from ancient viruses. Soon our RNA and DNA viruses are not just lowly parasites but are driving the evolution of large factories that can with ever-greater efficiency create more RNA and DNA cells. This leads to much greater survivability of the large cellular life, but the bottom line is that *the viruses are now hard-wired into the genomes of the cellular life. The diversification of cellular life, to exploit new energy sources and environments, is driven by the needs of the viruses—and might still be.*

This is a point that cannot be made too strongly. There was a primordial zoo going on here—an enormous variety of life and "not alive but on the road to" life, a huge variety of things. It is so easy to fall into the trap of envisioning one little cell finally making it ("I'm alive!") and then having it gradually diversify into more and more. Quite the opposite may have been true. There may have been a nearly simultaneous proliferation of life-forms of an enormous variety, all competing with one another in the primordial soup.

A recipe for making Terroans from RNA life

One of the proposals of this book is that life does not necessarily have to be enclosed by a cell wall to be life. In this section we continue in our cooking mode and see how cellular life could be concocted. The discussion will overtly refer to earth life, but by extension I suspect that this recipe will work for making non-Terroan cellular life as well. So take a look at what could be called the evolution of the cell, and for this account I summarize the magnificent work of Carl Woese, who has argued that one of the principal driving forces of cellular evolution was a process known as horizontal gene transfer (HGT), which comes about through the ease of swapping genetic code material between even distantly related lineages. When life was first forming on earth, whole segments of genetic code could jump from species to species, even when the species jumped to was in a whole different domain. So much genetic material was swapped early in life's history on Earth that some biologists have argued that even the great domains Archaea, Bacteria, and Eukarya are not lineages at all but amalgamations of jumping genes. HGT clearly had an effect on the evolution of life as we know it, and it would be naive to think that the same process would not also shape the evolution of alien life.

Early Terroan cells might have been like modular houses, with each part installed as a separate component. In its simple stages the separate components of the house—its heating, electricity, plumbing, insulation, entrances, windows, and roof—are readily exchanged. But sooner or later the house becomes (evolves) ever larger, and some components just won't work. Our new mansion needs a far more elaborate ventilation and cooling duct system than our simple house does. At what point does this happen: when the exchange of components can no longer successfully occur because of the increasing complexity of the house as a whole, when its systems become so intertwined that a slap-bang installation of some other system just does not work? In other words, at what point does complexity take over and hard-wire everything to the point that horizontal gene transfer can no longer work? For cells, this level of organizational increase would eventually have made horizontal gene transfer no longer practical or even possible. When cells were made of simple modules, swapping was rampant. But there came a time, according to the model now envisioned by molecular biologists, when the cell systems went from ephemeral to permanent. This is the point that Woese calls the Darwinian threshold. It is the point at which species, in something approaching the modern sense, can be recognized. It appears when a threshold occurs bringing about a higher level of organization and, in so doing, increasing the fitness of the cell. Natural selection favors these more functionally complex, integrated cells, and they flourished at the expense of the simpler modular varieties. To Woese, this threshold change is the base of the tree of life. What we call bacteria got there first and thereby quit accepting genes from the nascent Archaea and Eukarya, which were still part of the gene-swapping pool. Because of this, the Bacteria are separated from the Archaea and Eukarya, which share more characteristics. I think this is the right concept but the wrong time. This might be the base of the tree of Terroans, but not the tree of life. Much more was going on lower down. If LUCA is the base of the tree of life, it is a rootless tree. The reality, in all probability, was a rich and diverse root system from which the visible subaerial parts that we see so abundantly today first sprang.

We are confronted by a new view of how life arose on Earth. Our old view is Darwinian and thus processional. Protocell A begot B, which gave rise to C, each an improvement on its ancestor. The new view is of a huge variety of life wannabes, for the evolution of a true cell requires enormous novelty that seems impossible from a single evolving lineage. Whereas we look at LUCA as a single first creature giving rise to the three domains, the new view sees LUCA as a small army of precells, all madly swapping genes. Some of these would become Terroans. Some were ribosans. Thus, for a living Terroan cell to finally evolve there must be horizontal gene transfer, and there must be a large number of varieties around to swap with. In this view, disparity (a measure of the variety of designs) decreases through time to reach the first working cells. This is the same argument that Steve Gould proposed for the Cambrian explosion, in which a huge diversity of animal types is winnowed out, with only the most successful surviving. But the big difference is that the Cambrian animals, while affecting the ultimate survival of winners and losers by competition and predation, are not swapping their genetic codes. The microbes are not so stable. It turns out that genes are also made up of modules, and these can be swapped within a genome. Also, bacteria can still have horizontal gene transfer and can also take up free DNA in their environment and incorporate it into their genomic DNA. Bacteria and viruses swap DNA as well. This idea was controversial for decades but was shown to be the case in the 1990s, when a resistant strain of cholera was formed this way and washed up on the shores of Peru, as recounted by Laurie Garrett in her 1994 book, *The Coming Plague.*

Modern Terroan cells are born when the radical changing of genes stops. This takes place when the organism is no longer a number of semiautonomous fiefdoms within a newly built castle but becomes an interconnected economy. According to Woese, arriving at this grade of organization is the most important event in all of evolutionary history. It marks the beginning of life as we know it on Earth. I should like to differ with the master. It may have been an early ecosystem packed with ribosans and Terroans, viruses and cells and protocells— RNA-protein organisms, RNA-DNA organisms, DNA-RNA-protein creatures, RNA viruses, DNA viruses, lipid protocells, protein protocells: All these huge menageries of the living and wannabe living and just accidentally created chemical mistakes in one thriving, messy, competing ecosystem. Perhaps it was the time of life's greatest diversity on Earth, 3.9 to 4.0 billion years ago—the most important event in earth history.

A recipe for making DNA organisms from RNA organisms

How long did the RNA world last? Most workers suggest a relatively short time, but "relatively short" could be a hundred million years. Some people, including Steve Benner, think it may have lasted longer than that. In fact, we cannot falsify the idea that it exists still, with completely functioning (nonparasitic) RNA creatures hidden away on Earth, or having originated on, or been smuggled onto Mars, where they might exist still. That point taken, how to go to DNA from RNA? There might be two possible pathways. First, RNA gives up its dual role as informational molecule and catalyst and becomes solely a catalyst with newly evolved DNA as genome. Second, RNA gives up its catalyst role and becomes solely an informational molecule with protein as catalyst.

The true pathway depends on the properties of RNA itself. How capable was it of catalyzing ever more radical chemical pathways made necessary by the increasing number of "jobs" (chemical reactions) required by life's trying to better its efficiency at doing what life does: finding energy and replicating itself? If RNA is (or was) capable of really radical new reactions, it looks as if its path was the second.

What about DNA? When did it first appear? Could it have appeared before proteins so that it co-opted the "brain" functions from RNA, relegating RNA to the "body" functions now fulfilled mainly by proteins? In this case, DNA would have appeared prior to proteins. There may have been early organisms where both RNA and DNA stored information, with RNA acting to replicate DNA and to serve as catalyst in other reactions. In such an organism, still unnamed, RNA would also have taken over repair functions: RNA would recognize DNA and repair errors arising during replication. We can even envision newly evolved RNA-DNA creatures competing in, and easily winning, a battle with pure RNA organisms for resources simply by better replication of genomes. RNA bugs would have been very error-prone in this, to their sorrow.

The relative stability of RNA and DNA leads to fascinating conjecture. Biochemists routinely insult these molecules through a variety of tortures that would lead to intense picketing by animal rights crazies if these molecules were animals. But from this biochemical torture we know the relative stability of RNA when it is bound to another RNA in a double helix and of DNA in its double helix. Under most conditions, the DNA molecule is more stable, and the instability of RNA in an RNA organism would have led to common information loss, and death, during genome replication. The exceptions to this are in environments that are strongly acidic—here RNA does better—and in low-temperature, high-salt concentration environments. *This is just the sort of environment that we suspect to be present at the bottom of the Europan Ocean,* as we shall see in chapter 11.

We can thus see a possible progression of the RNA world, either from RNA with dual function to RNA/protein or to DNA/RNA. Why not both? Here might be a way that life, as we do not know it, alien life, could come about. A protein takeover of the catalytic functions of RNA makes a great deal of sense chemically and helps in the separation of genetic material from catalytic material, which in itself might be advantageous in avoiding replication mistakes. Proteins taking over the role of ribozymes is also more energy-efficient. In such a bug the ribozyme composed entirely of RNA would be replaced by a ribosome, composed of RNA and, at first, very simple proteins.

A final and fascinating possibility is that DNA came first. Could this have been the pathway, with pre-RNA skipping the RNA world altogether and jumping directly to the DNA world? This too could have taken place (although it is a less favored route, as mentioned earlier). The idea that DNA might have been more abundant in the "prebiotic soup" was first raised in 1967 by Woese on the basis of understanding how much more stable DNA is compared with RNA, especially in warmer environments.

This careful look at the deep past helps us understand the pathway to us Terroans. But what the various authors who have done this work are only just realizing is that we now have a rough idea of what alternative pathways might have occurred on planets and moons where initial starting conditions were not the same as on the early earth. The pioneers looking at the rise of earth life have also produced a series of models applicable to Mars, Europa, and Titan.

Relating the earth history to alien life history

Earlier I made reference to the Cambrian explosion. Here I shall invoke that short but important time in the history of life (and especially in the history of us animals) for two reasons. First, it reminds us how idiosyncratic any history can be. In *Rare Earth,* Don Brownlee and I devoted an entire chapter to the Cambrian explosion for some very specific reasons. Because that book dealt with our proposed rarity of the animal grade of evolution in the universe, we had to examine how animals came about on Earth. But there was a more specific question asked: Does any planet advancing to the complicated, animal-like grade of organization necessary go through a Cambrian explosion equivalent, in which most or all of the body plans (the phyla) simultaneously appear, or might there be other paths, where over long periods of time the higher taxa appear one by one? As a corollary we wondered why there were about thirty-five animal phyla involved—not ten, or a hundred, or any other number. These questions are still not answerable. But numbers may not always be by chance, just as the fact that there are two strands to DNA, instead of five, has very good and explainable chemical reasons. So invoking the Cambrian explosion is very relevant to looking at the evolution of the first cellular life, especially in light of this sentence at the end of Woese's review on how cells evolved: "Extant life on earth is descended not from one, but from three distinctly different cell types. [I think that this is a woeful underestimate, obviously, from the earlier discussion.] However, the designs of the three have developed and matured, in a communal fashion, along with those many other designs that along the way became extinct." So my question here: Why three? Could there have been five? Or one? Or twenty, leading to the current life on Earth? And how will the process go on some other planet? Was our history typical or just typical of an earthlike planet?

The *where* of life's origin on Earth: What kind of oven to cook in?

Why should we care about how life started on Earth in a book dealing with aliens? It seems to me that it might be likely that life could live on any number of places in space, including in our solar system. But living somewhere and coming to life may require quite different circumstances. If we want to understand the frequency of aliens, or the chances of finding aliens on Mars, for instance, we had better find out if there was any environment on Mars that could have spawned life. The kind of kitchen where life was first cooked on earth is certainly central in understanding the frequency of life in the cosmos. We can finish this chapter and its cooking metaphor by asking what kind of oven was used to cook up life? Here are some candidates.

A warm pond

The first, most famous, and longest-accepted model for life's appearance on earth was proposed by Charles Darwin, who in an 1871 letter to a friend suggested that life began in some sort of "shallow, sun-warmed pond," and as late as the 1970s, when those deep-vent expeditions were taking place, this was still the favored hypothesis. To this day this type of environment, be it of freshwater or perhaps in a tide pool at the edge of the sea, still remains a viable candidate in some circles and in textbooks. Other scientists early in the twentieth century, such as John Haldane and Alexander Oparin, agreed with Darwin and expanded on this idea. They independently hypothesized that the early earth had a "reducing" atmosphere (one that produces chemical reactions the opposite of oxidation; in such an environment iron would never rust). The atmosphere at that time may have been filled with methane and ammonia, forming an ideal (since it was filled with the chemicals necessary to create amino acids) "primordial soup," from which the first life appeared in some shallow body of water. Until the 1950s and into the 1960s it was thus believed that the early earth's atmosphere, thought to have consisted of methane and ammonia, would have allowed commonplace inorganic synthesis of the organic building blocks called amino acids by the mere addition of water and energy. All that was needed was a convenient place to accumulate all the various chemicals. Seemingly the best place to do this was in a shallow, fetid pond or a wavewashed tide pool at the edge of a shallow, warm sea.

Is this idea still credible? To many scientists now looking at environments on the early earth, the answer is a resounding "no!" Robert Shapiro, in *Origins,* summarizes the theory as follows: (1) The earth at the time of life's first formation had a reducing atmosphere with methane, ammonia, hydrogen, and water, but no oxygen and thus was strongly reducing; (2) this atmosphere was exposed to such energy sources as lightning, solar radiation, and volcanic heat, which, when combined with the atmosphere, led to the formation of simple organic compounds; (3) these compounds accumulated until the oceans reached the consistency of hot dilute soup, and from this came the now-much-used phrase *prebiotic soup*; (4) by further transformation life formed in the soup. The order of appearance was then simple protocells first, enzymes second, genes much later. For this to work, there must be methane in the early earth atmosphere. But even if methane was present, there is little agreement now that the overall chemistry of the atmosphere was reducing. The prebiotic soup fares no better. The organic compounds necessary to form life are complex and easily fall apart in heated solutions. Furthermore, an enormous amount of energy would be required to keep this soup out of equilibrium, which is necessary. Life will not form in soup.

As we learn more about the nature of our planet's early environments, tranquil ponds or tide pools—and primordial soup, even really dilute soup—seem less and less likely to be a plausible route and site for the first life or even to have existed at all on the surface of the early earth. What Darwin could not appreciate in his time (or Haldane and Oparin in theirs, for that matter) was that the mechanisms leading to accretion of the earth (and other terrestrial planets) produced a world that, early in its history, was harsh and poisonous, a place that was very far removed from the idyllic tide pool or pond envisioned in the nineteenth and early twentieth centuries, but that did not have a reducing atmosphere or ocean. There is currently a very different view of the nature of the early earth's atmosphere and chemistry. It is now widely believed among planetary scientists that the overall conditions would not have favored synthesis of organic molecules on the earth's surface.

The seashore

This is a variant on the warm pond but differs in some significant ways. It has been proposed that tidal pools would be a better place for life's beginning than warm ponds because a tidal pool offers some of the same operations that an organic chemist might use in his lab. Tidal pools have repeated drying and rehydration, they have lots of energy from waves producing bubbles of organic scum that might serve as protocells, and there is a complex chemistry going on with various salts and other chemicals that can be distilled and concentrated by the action of hot sun and tidal action in the myriad microenvironments of a tidal system with freshwater also flowing in from the land. This is a much more deluxe oven than the warm pond model.

Hydrothermal vent community

But if not in a pond or tide pool, where could the various components necessary for living cells come together to produce life? How about the absolute most deluxe oven on planet Earth! With the *Alvin* dives, described earlier, a new possibility was raised: Life on earth began in the newly discovered deep-sea vents. Soon new molecular techniques used to classify the vent microbes added confirmatory information to this idea. Most of the microbes from the vents were eventually found to belong to the Domain Archaea, microbes described in chapter 3. The Archaeans, it seems, belonged to the most ancient lineage of organisms known on Earth. But where would they have first come to life? While Darwin championed his nice warm little pond, we know that the early earth was a place constantly bombarded by giant comets and asteroids, making small ponds unlivable most of the time. The only places that would be insulated from the titanic energies would be the deep oceans, and in them perhaps only the hydrothermal vent system would provide the bomb shelters necessary for life's survival. But the problem is in making DNA and RNA in such settings. The energetic hydrothermal rift systems might not be conducive to these crucial components.

Cloud origin

A fourth locality where life might start on a planet like Earth, and perhaps on planets or moons not like Earth, is in the clouds, according to the man who gave us the tree of life. The research leading to the tree of life produced by the brilliant but irascible Carl Woese has changed our view of how life diversified. But what about its first evolution? Here too Woese has weighed in. Like Robert Shapiro, he is particularly disparaging in his view of the Oparin/Haldane hypothesis ("stultifying"). His own proposal is especially interesting because it deals with life forming in clouds. Since this might be the only mechanism leading to life on the gas giants such as Jupiter and Saturn, the Woese idea has much astrobiological relevance.

Woese puts the formation of life on Earth at a time earlier than in any other proposal. He envisions life starting even before the earth was fully formed and differentiated into its core, mantle, and crust components that we see today. In those early times there would have been large amounts of metallic iron present on the surface of the earth in contact with steam and some liquid water, amid an atmosphere filled with carbon dioxide and hydrogen. It is the last that is so interesting, since hydrogen is a potent driver of chemical reactions, but because of its light weight, it is easily lost to space on small-mass planets like Earth, Mars, and Venus (the gas giants are so massive that they can hang on to their hydrogen). At this time our planet was being barraged by space debris large and small, causing it to be encircled by a haze of dust particles and water vapor. High clouds of water vapor would form, and Woese envisioned that these droplets would have served as protocells, tiny cell-like objects. With sunlight as an energy source and the dust thrown up from the surface carrying organic molecules among the many other molecules and elements blasted into the sky by the asteroid bombardment, there would have been plenty of raw materials to make life from. With lots of hydrogen present as well, the first primitive organisms to evolve would have produced methane after using carbon dioxide as a carbon source. Microbes using this pathway today, hydrogen for energy and CO_2 for carbon, are called methanogens. As the earth cooled, oceans formed, and life fell from the sky to populate the oceans.

The cloud origin hypothesis was originated to explain earth life. But it might more appropriately be used as a model to understand how life might arise on very un-earthlike planets, ranging from Venus to the gas giants such as Jupiter, Saturn, Neptune, and Uranus that make up our outer solar system. And we know that the most common extrasolar planets that can be detected are Jupiter life worlds. Any mechanism that might lead to the formation of life in the upper atmosphere of extrasolar planets should be investigated.

So perhaps life could start in clouds. But would it be a specific type of life that we find over and over again, a kind of life that arises because of this particular mechanism?

On the surface of rocks

One of the most ingenious suggestions about how life did arise on Earth, and thus how it might arise elsewhere, comes from a German patent lawyer named Gunther Wachtershauser, who has proposed that the first life formed on crystals of a sulfide mineral called pyrrhotite, which is made up of iron and sulfur in crystal form. When this mineral is oxidized, it turns into the common mineral pyrite, which is also exclusively iron and sulfur but with a different crystal structure. According to Wachtershauser, this chemical reaction could have fueled an early life-form that sat on the crystals. This energy source and a primitive life that might arise on the faces of the crystals themselves are now called the iron-sulfur world hypothesis. Since life requires that a large number of organic (carbon-containing) molecules in a wide variety be available, and since in most cases many of these organic molecules have to be synthesized from simpler compounds, a powerful energy source is very important. Because both pyrrhotite and pyrite are found in abundance in hydrothermal vent systems, this energy system would not have been rare on the early earth—or on other bodies in the solar system either. What is somewhat amazing about this conception is that any life forming in this way would be tethered to or, more accurately, smeared onto the rock crystals themselves. Thus we can imagine some sort of essentially two-dimensional organism that has a powerful energy source available but that cannot reproduce. The next step would be the formation of RNA, with the final evolutionary step of this new life leaving the mineral surface to become autonomous. A brilliant conception, and such life would definitely qualify as life as we do not know it.

A newly proposed origin: Linked impact craters

It seems as if almost every conceivable environment has been touted for the site of life's origin on Earth: Darwin's warm little pond, hot springs, hydrothermal vents, waves, clouds, shallow seas, tide pools, almost everywhere but Steve Benner's choice, deserts. Suggesting a desert origin may be novel, yet this is just what Benner and his group proffer. They talk of a hot climate in alkaline conditions, where evaporitic minerals of the borate group would be found. Because of the poisonous effects of water on the prebiotic synthesis of complex organic molecules, such as RNA and many proteins, a lack of water, or at least a reduction of it, would be favored. Under such desertlike conditions, where the overall environment is alkaline and has calcium carbonate in abundance, the formation of ribose among borate minerals might be favored. Clay minerals of various kinds are also common in such settings, and increasingly it looks as if templates formed from clay would help bring about the synthesis of the complex organic compounds necessary for life.

Deserts have little water. Under such conditions another necessary ingredient for life—peptides, which are composed of amino acids—can also form and exist for some period of time. But life does need a solvent, and if not water, what? Benner thinks that a liquid called formamide might do the trick. A mixture of formamide and water in a hot pool under the desert sun would cause water to evaporate away, leaving a concentrated pool of formamide. In this mixture the formation of peptides from amino acids and RNA from nucleotides could take place. This is the stuff of life. Could a primitive RNA life have formed under these conditions?

There must be many "next steps," such as getting back into water somehow. If an RNA life was produced with a membrane that allowed its fragile RNA genome to be protected from the dissolving power of water, might we have the first life? This scenario seems as good as anything otherwise proposed and gets around the seemingly intractable problem of getting to RNA. It still has its failings, however, and yet another member of my team, Joe Kirschvink of Cal Tech, genius and one of my oldest friends, may have suggested the final piece of the puzzle.

Let us look in a little more detail at this desert origin of life idea. There are some fascinating ramifications, the most startling being that earth life did not start on Earth but on Mars.

For the borate mineral pathway to RNA to work, there has to be a liquid system that repeatedly decants and distills the liquids. Borate is an evaporitic mineral, forming only when liquid water evaporates away. One view is to have a single pond, or a desert oasis, that dries and refills with water in some fashion. But a better way to have this particular pathway succeed, the way that a good organic chemist would do it, would be in a series of distillation tubes, connected one to another. In fact, this might be the only way that it could be done. Since good glassware assembled by good chemists with good Bunsen burners were in short supply on the early earth, we need to come up with a setting where this pathway might succeed. To the rescue: Joe Kirschvink and his former student Ben Weiss, now at MIT, a pair we shall meet again in chapter 7. The two have come up with a natural setting that could lead to the formation of RNA from borate in the rough fashion that Steve Benner has suggested. But Kirschvink and Weiss have elaborated on the idea. What is needed, they say, is a system mimicking the chemist's bench, where a series of distillates are moved from container to container, allowing a more refined product in each successive flask. This is far more complicated than one small pond evaporating and refilling under the desert sun. Kirschvink invokes a system something like Mono Lake in California, where a series of lakes from higher to lower elevation have a linked flow of groundwater. The most obvious candidate for such a system, according to Kirschvink and Weiss, would be a series of linked impact craters in a desert setting, stepped and communicating craters going from higher to lower elevation and with a linked water table. In this way the same series of distillations and decanting could be accomplished. So far so good, and novel too: an entirely new proposal about the site of life's first origin on Earth. But could there have been such a site on Earth four billion years ago, when we think all this early chemistry was taking place? No, say Kirschvink and Weiss. For such a setting we need to go to Mars.

A Martian setting for life's first formation, using the borate pathway hypothesized by Benner, but then passing through linked craters in a desert setting, is the message, and the evidence to support it comes from the geology of the early earth. All the earliest earth rocks appear to have been produced in a water setting. In fact, there is no good evidence of continents on Earth until less than 3 billion years ago—on a planet 4.6 billion years old. This sounds suspiciously like the abysmal movie *Waterworld* of several years ago, which imagined planet Earth completely covered by water. Perhaps right vision, but wrong time. Kirschvink and Weiss can cite a lot of evidence supporting their contention that the earth, at the time when life would have first formed, had, at most, islands. Perhaps the linked craters could have formed on some island, but to Kirschvink and Weiss there is an easier solution. Mars never had planet-covering oceans, we are quite sure. Large lakes, maybe small seas—but huge oceans? No. The necessary desert could easily have been on Mars, but only with difficulty could there have been such a desert on Earth 4 billion years ago. The last trick was to take the first life and then get it to Earth. Not so difficult a journey, either, as we shall see in chapter 7.

How can we test this idea? Only by going to Mars. But one part of the whole RNA from borax idea can be done here on Earth. It turns out the some of the planet's mad Frankensteins are madly producing RNA right now in test tubes.

From cookbook to Frankenstein

In the last two chapters we have looked at plausible (and implausible) kinds of life and at the way that some of these life-forms could have evolved from nonliving to living. Up to now all these life-forms have been ones that we know on Earth or might find in the cosmos. But there is more than one way to make life. We humans are busily tampering with the origin of life in our laboratories, with the goal of making artificially created life. Some of this is Terroan in its characteristics, but some is not. The progress toward artificial life and the aliens that might be created is the topic of the next chapter.

Chapter 5

The Artificial Synthesis of Life

Indeed, biology may be said to be the sub-field within chemistry that deals with chemical systems capable of Darwinian evolution.

—A. Ricard and Steven Benner

The Frankenstein myth runs deeply and powerfully through our culture. It is about a culture-crossing taboo, the sin of creating life in some "unnatural" fashion, an action seemingly at cross-purposes with nature and creator. Yet genetic manipulation routinely creates microbes that were never evolved through natural selection and thus are as "unnatural" as Frankenstein's monster itself. But there remains a difference between these new bacteria and the hoped-for products of a small group of determined scientists. Whereas the new microbes are simply previously known forms whose genetic code has been slightly scrambled, the new efforts aim toward creating something altogether novel, something that may indeed qualify as an alien. How close that day is was recently illustrated by the eye-catching cover headline and its subhead on the June 2004 issue of *Discover* magazine: "What came before DNA? The startling attempt to creating living organisms in a lab." This is somewhat confusing, because there are two very different topics here. What came before DNA is indeed a profound and immensely interesting question and may (or may not) have anything to do with the artificial creation of life, since the easiest life to produce artificially might not be based on DNA at all. DNA, in fact, might be the Cadillac of information-carrying molecules, when a far simpler Ford Pinto version might do. Nevertheless, the article by veteran science writer Carl Zimmer was an eye-opener, for more reasons than one. For the first time an actual price tag was placed on the cost of producing artificially spawned life; chemists Jack Szostak and Steven Benner estimated that for a mere $20 million an artificially created life-form based on DNA could be constructed. It appears that the race is on, and it is likely that any such creation will qualify as an alien. In this chapter we will look at these enterprises, research chasing one of the greatest scientific quests of all, the creation of life in a test tube.

The classic Miller-Urey experiment

Our definition of life (at least in its simplest sense) can guide us: Life metabolizes, life replicates, and life evolves. We thus need machinery for the acquisition of raw materials (for building material and for internal liquid), for the acquisition of energy, for the production of copies of ourselves, for the production of, or replacement of, the material that we are built from, and an information storage system. All these functions are done in a well-known fashion by earth life. Changing any or all would qualify our new bug as an alien. But all are needed if our new bug is to be alive. Where are we in creating such new life?

The first attempts even in a crude sense toward creating life in the laboratory began at about the same time that James Watson and Francis Crick were seeking the molecule that was ultimately named DNA. In those post–World War II years the scientific community was examining what kind of environment, materials, and energy would be needed to allow life to originate. Then, in a breakthrough experiment, it became clear to all that the building blocks of life had probably been present in quantity on the early earth. In 1953, Stanley L. Miller and Harold C. Urey, working at the University of

Chicago, took gas molecules, which were believed to represent the major constituents of the early earth's atmosphere, and put them into a large glass flask partially filled with water. The gases were methane, ammonia, hydrogen, and water vapor. Next, a continuous electric current was discharged in the flask, to simulate lightning storms believed to be common on the early earth. At the end of one week Miller and Urey observed that as much as 10 to 15 percent of the carbon had formed organic compounds. Two percent of the carbon had formed some of the amino acids that are used to make proteins. Perhaps most important, the experiment showed that organic compounds such as amino acids, that are essential to cellular life, could be made easily under the conditions that scientists thought were present on the early earth. This finding inspired a multitude of further experiments and lent credence to the idea that life was started inorganically and that the process was not all that difficult.

Surprisingly, this auspicious and promising start was followed by nearly a half century of frustration. While it was clear that the building blocks of proteins could be created in the lab, the more complex molecules, such as DNA and RNA, were much harder to produce. Only in the late twentieth century and the first years of the twenty-first has progress on this front continued in anything like a promising fashion. But progress there now is, and optimism too that the elusive quest of life's artificial creation might be within reach.

It is difficult to underestimate the influence of the Miller-Urey experiment, but recently there have been astute critics—namely, Robert Shapiro, in his book *Origins,* which should be required reading for anyone interested in the origin of life. Shapiro should have named his book *Skepticism*—not of course skepticism that life originated but that so many of the tropes about the process have received so little critical attention. The Miller-Urey experiment is one area that Shapiro believes has been overrated. At the time of the experiment it was thought that the earth's atmosphere was a reducing one; we now think that if it was reducing, it was barely so. Our idea of what gases were present has changed as well. Finally, if we acknowledge that earth life is formed from lipids, proteins, carbohydrates, and nucleic acids, is it not telling that the Miller-Urey experiment produced none of these? The products produced are vastly different from the parts of any living cell.

There are two parallel fronts in the quest to create life, and they mimic two great and opposing ideas about how life arose. Oparin-Haldane-Dyson thought that proteins and then protocells came first, and the information storage molecules only later. Woese and others see information first, in the form of nucleic acids, followed later by cells and proteins. Today one group of experimenters is trying to make artificial cells, which they hope to inoculate with naturally formed RNA cribbed from a living organism. Another group is trying to see how RNA was produced in the first place.

To build DNA requires energy, amino acids, chemical concentration, catalysts, and protection from strong radiation or excess heat. Our modern-day Dr. Frankensteins have to complete four steps:

1. The synthesis and accumulation of small organic molecules, such as amino acids and molecules called nucleotides. The accumulation of chemicals called phosphates (one of the common ingredients in plant fertilizer) would be an important requirement since these are the backbone of DNA and RNA.
2. The joining of these small molecules into larger molecules, such as proteins and nucleic acids.
3. The aggregation of the proteins and nucleic acids into droplets, which take on chemical characteristics different from their surrounding environment.
4. The ability to replicate the larger complex molecules and establish heredity.

While some of the steps leading to the synthesis of RNA and the even more difficult feat of producing DNA can be duplicated in the laboratory, until recently others could not be. There is no problem in creating amino acids, life's most basic building block, in test tubes, as shown by Miller-Urey. But it has turned out that making amino acids in the lab is trivial compared with the far more difficult proposition of creating DNA artificially. The problem is that complex molecules such as DNA (or RNA) cannot simply be assembled in a glass jar by combining various chemicals. Such organic molecules tend to break down when heated, suggesting that their first formation must have taken place in an environment with moderate, rather than hot, temperatures.

Making life

Although most analysis of what we might call making life in a test tube, or the creation of artificial life, centers on the difficulties of creating the information-containing molecules, the wall of the cell itself is no less important and, according to one school of thought, predates the evolution of either DNA or RNA. The problems are not trivial. Most vexing and necessary is the coupling of processes between the border of the cell and the functions of the interior.

The formation of the wall membrane is only one aspect of artificially reproducing an autonomous living cell. While there have been numerous attempts to illustrate this process computationally, the artificial synthesis or in vitro systems seem to have the most promise. Such systems hold much promise too for the formation of biotechnologies and production of pharmaceuticals. To those interested in less profitable ventures, building artificial cells also furthers our understanding of earth life's first evolution, as well as opens the door to the formation of artificial and potentially alien life. This research has been called the minimal living cell program.

Research into the formation of a minimal living cell comes from two directions, which we can call the bottom-up and top-down approaches. In the former, individual chemicals and components are mixed together in an attempt to synthesize a living cell. In the latter, an already living cell is parsed of various components (without killing it) by removing various bits of its matter.

It is the bottom-up approach that seems to harbor the greater chance of creating artificial life that qualifies as life as we do not know it, or alien life. The real challenge for it is to create chemical automata using components other than those found in biological systems known to us. This could also give the key to part of the problem of the origin of life on Earth, since the precursor molecules could be different from the biomolecules that later came to integrate full-fledged living beings.

One of the key clues to understanding life's evolution came not from the earth but from space. During the analysis of a class of meteorites called carbonaceous chondrites, mineralogists and organic chemists studying these rocks that have fallen to Earth from space were astounded to discover the presence of organic compounds, including amino acids, as well as simple amphiphilic molecules. If they fall to Earth now, they must have fallen to Earth early in our planet's history, and it was quickly surmised that the material needed for proteins—amino acids—as well as raw materials for primitive cell wall membranes were available on the earliest earth, before life started. The latter, the amphiphilic molecules (*amphiphilic* means "loving both sides," a good term for something that must deal with two different environments, or for a molecule composed of two layers), are of particular interest. Cellular life found today uses a complex lipid (fat) bilayer, but in the earliest phases of evolution, cell layers must have been (and could have been) far simpler. There are many candidates for these first cell wall membranes, and today these can be formed by a variety of common compounds, such as soap molecules, glycerol monooleate, oxidized cholesterol, and even some detergents.

Any castle can be won if one simply starves out the inhabitants. So too with the early castlelike walls created in protocell experiments. If the wall membrane is too efficient at keeping things out, the inhabitants starve. Castles need a constant ingress of food, water, and other materials necessary for life, and one of the conundrums in origin of life scenarios has been to try to build artificial cell walls that can keep in the good stuff (genes and protein machinery) while keeping out the riffraff (toxic chemicals) and at the same time let-

ting in food, energy, and liquid. Cell wall membranes are complicated, and their means of letting ions and molecules in and out is
through a series of chemically complicated channels. Enzyme transport systems that move ionic nutrients and metabolites across the
cell layers are highly evolved and highly sophisticated. So how did
primitive life accomplish this with much simpler systems, and how
can our modern-day Dr. Frankenstein do the same?

Lipid bilayers make marvelous wall membranes, but they are
nearly impenetrable to ionic solutes (like the components of dissolved salt and sugar) and bipolar molecules, like water. While much
effort has gone into producing working genomes, a cell, like a castle,
is nothing without its protective wall. How did these walls come
about in evolution, and what is the progress in making them in the
test tube environment?

This question has been on the minds of those who would construct artificial cells. Central to the formation of synthetic membranes has been David Deamer and his colleagues. Deamer and his
group have shown that protocells are easily made in the lab and
hence in nature. Even something as simple as mixing gasoline and
water can make tiny protocells, while the use of fats and water, or
glycerine and water, can also do the job. There appears to have been
a multiplicity of protocells easily made on the early earth and perhaps on other planets as well. This research suggests that the con-
tainers for genomes were not the most difficult step in forming life.

Protocells and bottom-up engineering

The two methods of artificially creating life already mentioned are
bottom-up, essentially starting from scratch and synthesizing all the
components of a cell (its wall, genome, and metabolic and replication machinery), and top-down, taking an existing cell and simpli-
fying it by removing parts of its genome to minimal levels.

The bottom-up group is going at it from two directions. While one group of Dr. Frankensteins is busy figuring out how to make RNA and DNA, another and very promising direction comes from those trying to make small cells first. The idea that protocells (without genetic machinery) came first in life's evolution was the brainchild of the late Sydney Fox. In this view, tiny compartments with walls formed first, and only later (and in some still-mysterious fashion) were the cellular machinery and genome somehow inserted. Today many groups are following this lead. Some of the most promising work in this direction has come from a group from the Los Alamos National Laboratory headed by Steen Rasmusson and by another group including David Deamer and colleagues.

At a meeting in 2004, Rasmusson and his colleagues presented their progress on what they deemed a minimal protocell design, in which a small aggregate of fatlike substance (actually lipid) acts as a tiny container. Within the tiny space a PNA molecule is inserted. PNA is an analog of DNA. It is an information storage molecule with a backbone not of sugar, as in DNA and RNA, but of a material called a pseudopeptide, which is akin to a protein. These tiny protocells are surprising: They have metabolism, with light as an energy source, and the material they produce is lipid and more PNA. This exciting development approaches life in some of its characteristics. The so-called Chen-Rasmusson protocell for the first time integrates genetics and metabolism in the same small containment vessel. Is it alive? Not yet. The system can neither replicate satisfactorily nor evolve.

Top-down monsters

The target organism for the top-down experimentalists is the charming microbe *Mycoplasma genitalium,* a bacterium that does profoundly disgusting things to the urinary tracts of humans. It turns out to be the organism with the simplest-known genetic code, having only 580,000 bases in its genome, coding for about 480 genes. The experimentalists take this bug, extract some of the genome, and add new parts. It has already been determined that at least 130 of the 480 genes are not essential for life, and another 85 that may not be essential either have been identified. One of the most active participants in this work is Craig Venter, who helped decode the human genome. A large group that includes Venter is working with *M. genitalium,* with the espoused purpose of creating bacteria with specific activities encoded with artificial genomes. In this case use of the word *bacteria* may be informal, for if the enclosed genome were sufficiently different, our microbe would no longer be a bacterium, one of the three great domains of life.

A synthetic virus

While much of the work dedicated to creating new life is concentrating on the construction of a bacterium-like organism, other groups have concentrated on creating viruses or parts thereof and have succeeded. If we consider that a virus is a life-form, then artificial life has already been created.

Two groups have been prominent. One group, headed by Eckard Wimmer of Stony Brook, successfully synthesized the genome of an ancient enemy of humankind, the poliovirus. The group went on to build an infectious particle with a genome length of seventy-five hundred nucleotides. Was this particle alive? The journal *Nature* (which, along with *Science* magazine, constitutes the flagship of scientific publishing) had no doubts. It dedicated an editorial to the subject of synthetic life in general and the Wimmer results in particular, stating that "the synthesis of the poliovirus capable of replication, *and therefore a life form* [my emphasis], can be put together by laboratory procedures without the intervention of any native viral DNA." If we are to accept *Nature*'s conclusion that this infection particle is indeed alive, we see here the synthesis of life.

This is not the only virus synthesized in recent months. The omnipresent Venter and crew constructed an entirely synthetic genome for another virus, a parasite on the common human-inhabiting bacterium *E. coli*. Once again the new viral particle was infectious, demonstrating traits that many would equate with life. The genome of this new viral Frankenstein's monster was slightly shorter than that built for the poliovirus, but impressive nevertheless, almost fifty-four hundred nucleotides in length. What was even more impressive was the rapidity with which it was synthesized; it took only fourteen days to build this genome from scratch. While that is far longer than it takes nature to do this work, it certainly beats the unknown number of years that it probably took to produce the first life on Earth.

These two examples show that the methodology for synthesizing the genome of even the most complicated virus is now available. The same methodology can be used to build a bacterial cell as well or at least something that resembles a bacterium, a form of artificial life larger than a virus. The reaction to these two triumphs was quick in coming. While most were congratulatory, there was also a visceral negative reaction from part of the scientific community. It was pointed out that in this day of rampant terrorism, the same methodology could be used to produce truly terrifying viruses that could be used as weapons against large population centers. Perhaps in a further reaction, some in the scientific community began to reflect on the implications of these godlike acts. When Mary Shelley wrote her now immortal *Frankenstein*, the world was a far more biologically innocent place, but even so, the ethical considerations of synthetic life were troubling. Is it still so? An ethical review panel, commissioned to consider the implications of the Venter virus, found no evident problem with this new line of work, concluding: "The prospect of constructing minimal and new genomes does not violate any fundamental moral precepts or boundaries, but does raise questions that are essential to consider before the technology advances further." Later we shall revisit some of these implications and the potential for mischief that this new technology could spring upon humankind.

Synthesizing RNA life

The man who has perhaps worked hardest at synthesizing life is Jack Szostak of Harvard. He is quite clear about what to do and how to do it, and the following discussion is adapted from one of his articles on this state of progress. Szostak is bullish on making life. To him, it is not a question of if, but of how and when. His optimism comes from his success in coaxing the process of natural selection to work for his experiments. Starting from scratch and getting to life in a single experiment may be impossible. But beginning from a midpoint and then using natural selection to home in on a living cell are foreseeable.

Szostak sees the starting point in the life synthesis process as attaining a collection of molecules simple enough to form self-assembly yet sufficiently complex to take on the essential properties of life. He believes that the living cell eventually produced will have followed in the footsteps of life following the RNA world hypothesis. He puts it this way: "We believe that within this framework structures can be found that are both indisputably alive and yet simple enough to be amenable to total synthesis." His definition of life is the same as that which we discussed earlier: "simple cellular systems that are both autonomously replicating and subject to Darwinian evolution," a system that is autonomous, replicates, and evolves, something that shows growth and division and that is reliant on the input of small molecules and energy and thus metabolizes, another of the required elements in our definition of life.

Szostak and his Harvard team regard synthesized RNA as the key component of this synthesized life. RNA will act both as the template for the storage and transmission of genetic material and as the catalyst that will aid in replicating itself. It is this RNA polymerase, a term for a catalyst, that will replicate RNA. But a single strand of RNA by itself cannot be considered alive. To work at all, it needs one strand to serve as a catalyst and a second strand to serve as a template. Acting together, these can start turning out new strands of

RNA—with their original code. If these two suitors are ever to meet, they need to be contained in something, instead of floating around in some organic soup for millions of years, hoping to hook up. Second, and more important, our starting RNA will be simple and probably replicate very inexactly. For life to work, RNAs will need to evolve into something more complicated and something that replicates with a very low error rate. By keeping molecules that are closely related together, advantageous mutations—errors in replication that lead to greater efficiency of either copying, or greater efficiency of catalyzing—can occur. What a marriage! Better and better replicators aided by better and better chemical catalysts to help put together the dear little baby RNAs. So compartmentalization goes hand in hand with the evolution of better RNA molecules.

Time goes on. We have small little compartments, protocells, with RNA of varying quality within. The protocells with better replication will replicate more efficiently; this is an evolutionary advantage. The protocells are made up of the lipid membranes that I have described earlier; this seems the simplest way to build a cell membrane. Yet the RNA does not yet command the protocell; they are roommates but still independent. The vesicle serving as the container must spontaneously replicate independent of the RNA within. The membranes interact with lipid molecules in whatever environment this is taking place in (a test tube today, the ocean or some liquid in the deep past), while the RNAs are doing their thing within. There would be a tremendous amount of variation in all this: small protocells, big protocells, and lots of RNA, little RNA, along with a great variability in the type and functioning of the RNA.

We have reached a point where a true living cell is in sight. To get there, our RNA must code for some activity that synthesizes the amphipathic lipid that makes up the protocell membrane. This would happen if the ribozyme within the protocell—the RNA mass that makes more RNA—starts making cell membrane material as well. With this, the membrane grows, and for the first time the RNA and the protocell are coupled, allowing natural selection to work on the whole shebang. As Szostak notes, "A simple cell with an interdependent genome and membrane would be a sustainable, autonomously replicating system, capable of Darwinian evolution. It would be truly alive." It would also be a living form that is not currently covered by our classification of life. It would be an alien.

So where are we in this deceptively simple process? Life, after all, had hundreds of millions of years, maybe a billion years, to go through these steps. Humans have been trying to make life in a test tube for no more than fifty and with any sort of sophistication for less than ten. What are the challenges ahead, and what progress has been made?

The history is somewhat abbreviated. The first challenge was to paste small RNA molecules together; a typical DNA genome of a very primitive bacterium is composed of hundreds of genes. It was soon found that there had to be some energetic driving force to convince the soup of short-strand RNAs to polymerize, or merge. Even in very high concentrations of an RNA "soup," few long-sequence RNAs were produced. Eventually RNA called a class 1 ligase ribozyme carried out the required polymerization, forming long-strand RNAs. Soon these chains were growing at the rate of one nucleotide (the rungs of the ladder, if it were DNA) per second. This process has advanced to the point that a true RNA replicase is, as Szostak notes, "tantalizingly close." We are on the verge of an RNA that can indeed catalyze the replication of another RNA, one with code.

We are at the point where we can soon start adding ingredients. The RNA polymerase is not yet good enough to replicate accurately the RNA, which serves as the code. At the present time too many errors arise. But Szostak and others pursuing this goal are closing in. Szostak estimates that a hundredfold increase in polymerization and a tenfold increase in replication fidelity will result in an RNA that can replicate itself with sufficient accuracy to stay alive. We're not there yet.

But the group has made progress. His is not a Johnny-come-lately search, for Szostak has been working on an artificial RNA molecule for more than two decades. His launch into this area of research was stimulated in part by the early eighties discovery by Thomas Cech that RNA could be both messenger and catalyst. Szostak, like any good Dr. Frankenstein, began to tinker with microbes, including the same organism studied by Cech and his colleagues, a bug named *Tetrahymena*. But like any good scientist, he needed the patience of Job, for this was not to be a project easily or quickly done (the putative record for creating life, in seven days according to a really old publication, may never be bettered, and there is some skepticism about the ability to replicate the methods used). His first work, with grad students Jennifer Doudna and Rachel Green in 1991, succeeded in making a very crude molecule that could take short hunks of RNA molecules in solution and make them copy themselves. He had artificially instigated the process of replication. This is far from life, but it is a start; he had created an artificial ribozyme. It could only replicate an RNA molecule several nucleotides long—well short of a single gene. And talk about poor workmanship; the ribozyme made numerous mistakes in copying. It was very, very crude.

So how to improve this first effort? Szostak and his students used one of nature's own tricks, natural selection. We have only to look at the domesticated vassals of humanity—the cows, sheep, domestic dogs, and so much else—to see how efficient a process of natural selection can be. All the domesticated species are derived from wild species with similar characteristics; all seem to show rapid growth to maturity, an ability to breed in captivity, little tendency to panic when startled, amenable and tractable dispositions, and social structures and hierarchies that favor domestication. A brutal form of natural selection then further selected for all these characters: Individuals that showed the favored characters were allowed to breed; those less favorable for human needs were killed. Highly sophisticated evolutionary products were rapidly produced through a simple method: Save those that taste good or cuddle (pets or meat, as Michael Moore once so unforgettably described white rabbits), and kill the rest. Szostak and his group have done this with their artificial RNA. They started building gibberish RNA by chemically stringing together nucleotides with their sugar backbone to make trillions of small RNA molecules. Each of these new molecules had a single task to perform: capturing and bonding to another molecule. But through elaborate chemical means the chemists could cull out the molecules that were capable of latching on to other molecules and discarded the rest. The winners were then required to make copies of themselves, and those that could were kept, while those that could not went down the drain of the lab sink, or wherever chemists discard poor rejected RNA molecules that cannot replicate accurately. Over and over this process was repeated, with a few winners and many losers in each replication of the experiment. But in this way natural selection (or unnatural selection, because the selectors were the humans), created new molecules. The winners were named aptamers, or parts that fit. Over time little aptamers were created that could bind to very specific things, some big: viruses and assorted proteins.

From this start—aptamers that can replicate themselves faithfully and can also stick to things—Szostak over the years increased their complexity and functions. Like a flea circus, they were "evolved" to do ever more complex feats, including the all-important trick of cutting DNA molecules in half. But the one trick that still eludes the lab is the most important: getting an aptamer to not reproduce itself but to take chemicals in the test tube and assemble them into a new and long RNA molecule.

How close are we to this milestone? The best to date is already more than ten years old, for in 1993 David Bartel, a grad student of Szostak's, made an aptamer that could stitch one RNA fragment to another. Now Szostak had done his own replication trick, turning Bartel into another Dr. Frankenstein. As of this writing, their captive aptamers can stitch together an RNA molecule with fourteen nucleotides added artificially to it. The next quantum leap will be a stitched-together nucleotide of one to two hundred nucleotides.

So at what point will a stitched-together RNA molecule, enclosed in some human-made protocell, have enough information to do what life does: replicate, metabolize, and evolve? No one knows, but what we do know is that it is coming. The major hurdles seem to be overcome, and now it is just a matter of time. That means that we had better make some room on the tree of life for this still-gestating bundle of joy to be delivered into the world by biochemist storks. (This is a major reason that I feel it necessary to do some tree work and one of the reasons I decided to write this book.)

The creation of new life

It is clear from this chapter that the dreams of Dr. Frankenstein are being realized. We have already created aliens, life as we do not know it and are about to make more—a whole tree-full. Where would such life fit into the classification scheme for earth life? Would these life-forms be Terroans? And what of life found elsewhere in the solar system?

Chapter 6

Are There Aliens Already on Earth?

Back before the first dinosaurs, before the first fishes, before the first worms, before the first plants, before the first fungi, before the first bacteria, there was an RNA world—probably somewhere around four billion years ago, soon after the beginning of planet Earth's existence. . . .

—Matt Ridley, *Genome*

The concept of extinction did not exist until the nineteenth century. Until then there was no sense that species evolved, lived for a time, and then went extinct. Because religion still held sway over so much of the world, there was a sense that everything God had ever created would still be found—somewhere. Even when large bones and shells of creatures clearly unlike those familiar to the noted naturalists of the day were uncovered from sedimentary strata, the belief was that the creatures in question were still alive somewhere in the vast, little-explored world. But by the early 1800s there was little of the world still unexplored, and it fell to Baron Georges Cuvier (struggling to keep his head on his shoulders through the French Revolution and its aftermath) to demonstrate the reality of extinction. Cuvier, the so-called father of comparative anatomy, obtained the skull and teeth of a proboscidean that was distinctly different from either the African or the Indian elephant.

He announced that the fossil elephant parts came from a species now extinct.

Today, in a world where the extinction rate among endangered species is occurring at an unknown but surely significant rate, it is pleasant to think of a world without extinction and one so little explored that even in the early 1900s Sir Arthur Conan Doyle could put his Lost World with its diverse saurian legions atop a large plateau in the Amazon region and readers could retain a sense of why not? Today we live in a world that is the opposite. There is a sense that it holds no more secrets, that the vast majority of its biota has been discovered and cataloged, that even the immense oceans, now steadily giving up their most famous wrecks to the Bob Ballards of the world, hold little mystery. But this smugness that nature and its secrets are now conquered might be a bit premature. Two recent events have made this abundantly clear. First, in 2002, oceanographer Debbie Kelly of the University of Washington (and another member of our astrobiology faculty) made a startling discovery deep beneath the Atlantic Ocean. Amid the mid-Atlantic Ridge, using submersibles and remote cameras, Kelly and her crew found their own Lost World and named it the Lost City. They found the equivalent of the black smokers profiled in chapter 1, but this time the substrate was white limestone rather than the darker peridotite rock of the Pacific Ocean smoker systems.

The structures discovered are spectacular. Some are nearly two hundred feet tall, growing like monstrous white stalactites up from the bottom of the sea. These chimneys vent a diverse chemistry of fluids with temperatures ranging from less than 40°C to over 90°C. The fluids coming out of the white chimneys are enriched in methane, hydrogen, and hydrocarbons other than methane and are thus rich sources of energy in form that can be utilized by life. And unlike the black smokers of the Pacific, the microbes living among and in these white chimneys are fueled by the chemistry of rock-altering reactions, not simply the heat coming from deep earth. The Lost City hydrothermal field is thus unlike any known submarine vent system and, according to its discoverers, may be our closest analog to early Earth

*Hydrothermal vent deposits taken in the Atlantic Ocean. The whiteness of
these deposits results from calcium carbonate minerals.* (Courtesy NOAA/ALVIN)

and early Mars, on the basis of our understanding of the chemistry on
those planets at the beginning of their history. The temperatures are
moderate, the fluids have a high pH, and there are low metal concentrations and high concentrations of energy sources. The Lost City,
which may be our best hope for finding a Lost World of early life on
earth, was unknown until 2002 and has been visited only once. There
is certainly an abundance of new microbes—and perhaps environments as well—yet to be discovered there.

It is not just the deep sea that is yielding new microbial discoveries. Soon after the discovery of the Lost City field, Craig Venter,

turned his gaze toward the sea. Venter sampled seawater from the Sargasso Sea of the Atlantic Ocean and, using the same powerful gene mapping techniques that were perfected in the Human Genome Project, surprised the biological community in identifying, from relatively small samples of seawater, 148 previously unknown types of bacteria. This was a shock to the complacent microbiologists who had thought that we pretty much had a handle on what was out there in the ocean. Furthermore, Venter identified 1.2 *million* new genes from these samples. These add to an already heady number of known Terroan species. Of the earth's currently defined creatures, about 750,000 are insects, 250,000 are plants, 123,000 are arthropods exclusive of insects, 50,000 are mollusks, and 41,000 are vertebrates, with the remainder composed of various invertebrate animals, bacteria, protists, fungi, and viruses, leading to a total of about 1.6 million named and identified species on the earth. The majority of these leave no fossil record. The precise figure is not known, since there is no central register for the names of organisms, and because of this, many species have been named several times. Taxonomist Nigel Stork points out that this level of synonymy may approach 20 percent. For example, the common "ten spotted ladybird" found in Europe has forty scientific names, even though it represents but a single species. Such mistakes are made because many species exhibit a wide range of variability. Often the more extreme examples of a given species are mistakenly described as new or separate species.

Does this mean that the number of species on Earth today is less than the currently defined 1.6 million? Probably not. Most biologists studying biodiversity suspect that the number is far more than this, but an intense debate ranges about exactly how many more. The most extreme estimates are in the range of 30 to 50 million species, meaning that taxonomists have named a bit over 3 percent of species on Earth and thus, in the 250 years or so since Linnaeus set out the task of describing every species on Earth, have barely begun their work. Other, more cautious souls posit a much lower number of between 5 and perhaps 15 million species. Yet even with this lower number it is clear that the work of describing the earth's fauna has a long way to go. There may be millions of new bacteria to discover and identify alone. We thus see that we may have recognized only a fraction of the total of earth organisms. Is the same true about the diversity of life chemistry on Earth as well as the diversity of species?

These discoveries demonstrate that indeed there is much yet to be learned about the diversity of life on Earth and more about its habitats and biochemistry. This is especially true for the Archaean domain. These microbes show a huge diversity of metabolic pathways just now being discovered. Just as there are many new microbes being discovered, so too was it a surprise how incomplete our knowledge is of something as basic as the fixation pathways of carbon dioxide by microbes on Earth. Where we thought that there were only a few ways that microbes could create food from the carbon in CO_2, it has become clear that we are only scratching the surface of what microbes can do. This gap in our understanding is obviously a huge hurdle for those pondering what might be encountered on the many diverse habitats that could be present in the solar system.

Here is a sense of how fast the discoveries are mounting and an indication of how much is yet to be discovered *and how much might still be hidden.* By 1987 microbiologists had identified twelve major phyla within the domain Bacteria, based on ribosomal RNA analyses. By November 2003 there were 52 phyla identified, and by 2004 the number had risen to more than 80. That is why, when Venter found 148 new types of bacteria, from surface seawater at that, it was a major surprise.

If there are so many new microbes as well as vast new features of the planet being discovered, what else is hidden from us? Might there be something even more spectacular out there waiting to be found, such as non-Terroan life? The proposition no longer seems so ridiculous. In fact, we might ask if a whole alternate biosphere exists on earth in tandem with our familiar DNA biosphere. This is really a dirty little secret. While we continue to affirm that there is but one form of life chemistry of Earth, in reality we do not know. Moreover, if there were a second dominion of life on Earth, what would it be? The most probable of such life as we do not know it would be RNA organisms, the group that I have defined as the cellular forms. As we have seen, many believe that cellular RNA life was the stepping-stone to modern-day DNA life but then went extinct. I certainly do. Yet strange living fossils, such as the coelacanth fish and the monoplacophoran *Neopilina* (a very early mollusk that was at the base of the tree of molluscan phylogeny), show that once in a while a living fossil comes out of the woodwork. Might the same be true of RNA cellular life? We still have RNA viral life. Why not the larger kind?

Both coelacanths and monoplacophorans were found by accident. These living fossils were fished out of the deep sea. But both are relatively large. The coelacanth, a bright blue fish with lobed fins, the supposed ancestor of us land-living vertebrates, is up to six feet long; the monoplacophoran is considerably smaller, only about a half inch in shell length but nevertheless something that can be seen without a microscope. For microbes, identifying something really new is a much more difficult proposition. A very large microscope is necessary to see even one, and unlike the fish and the mollusk, which can be easily recognized as radically different from the many other members of these tribes, almost all microbes look alike. It takes an inspection of their genes to recognize their degree of difference from other microbes. Herein lies a huge rub: It is not clear that we would recognize an RNA organism if we found one. We may have already found millions of them that remain unknown and unidentified, not recognized as non-DNA life. Why? Because the method we use to differentiate DNA life is by comparing DNA, or ribosomal RNA. We target the ribosome for study. But RNA life would not have ribosomes! Our RNA life would have a different kind of RNA, presumably, and thus would be invisible to our tests!

The potential presence of still-living RNA life on earth has intrigued biologists for some time. Microbiologist (and Nobel laureate) Joshua Lederberg has advanced ways that we could search for another possible non-DNA life on earth, protein life. His solution is to culture samples of microbes with radioactive phosphate. Terroans would incorporate this poisonous material into their nucleic acids and perish, whereas any possible protein life would survive. Another method would be to try to grow microorganisms in a medium without phosphate. Again, anything that grew would not be our familiar earth life. Lederberg has called on his colleagues to start a search for non-DNA life using substantial research funds, for the search would be neither trivial nor easily accomplished.

Unfortunately the search Lederberg suggests has never been attempted. That's a shame, for on much less evidence and chance of success, the SETI organization continues to look for intelligence among the stars when there is still so much mystery on the ground beneath our feet.

Life beyond Earth

This chapter has been short, as it should be. We know nothing of alien life on Earth (other than the aliens that we have constructed, of course), only that it might exist. And if this is unsatisfactory, what of the even larger question and larger unknown? What can be said of life beyond Earth, something almost universally suggested but never demonstrated? We will see in the next chapters as we leave the earth in search of alien life.

Chapter 7

Panspermia: Why There May Be Aliens Throughout the Solar System

Rather than worry about how life was created here on earth, it made more sense to think how it might be transported here from elsewhere.

— David Koerner and Simon LeVay, *Here Be Dragons*

L et us imagine a universe that is infinite in extent and age. This is a universe that did not start with a big bang. It never started at all, since it has always been here. In such a universe even the most remote possibilities become probabilities. In such a universe, in fact, we might expect life everywhere, moving from place to place through remote chance on celestial tramways: comets, asteroids, even rogue planets, carrying the spark of life from dead world to dead world. This notion of a steady state universe seems ludicrous to us now, but in the nineteenth century it was gospel. And in such a universe the transfer of life would be inevitable, or so believed the greatest scientific minds of the time, including Lord Kelvin and Svante Arrhenius. They named this hypothetical spreading of life throughout the universe panspermia. Today the idea is making a comeback, on the basis of new information. That life can and has traveled from Mars to Earth, or Earth to Venus, or from Titan to Europa is the center of a lively debate, and according to all who study the prospect, it is distinctly possible. In this chapter we will look at the profusion of new evidence both for and against the panspermia hypotheses and why it is central to our search for alien life.

ALH 84001 and the revival of panspermia

The idea that life could have come to Earth from space dates back to about 500 B.C., when the Greek philosopher Anaxagoras wrote about "the seeds of life" that spread through the cosmos. By the 1800s the concept that life might spread through space was seriously discussed first by Sales-Guyon de Montlivault, who thought that earth life had been seeded from the moon. Later the idea was explored by the great German physicist H. E. Richter, who first recognized that meteorites contain carbon and then made the intellectual leap to suppose that meteorites could have brought the first life to Earth. This concept was accepted and promoted by the most famous scientist of the nineteenth century, Lord Kelvin, who noted: "We must regard it probable in the highest degree that there are countless seed-bearing meteoric stones moving about through space."

At the end of the nineteenth century this idea was a hot topic in science, and in the early twentieth century, its main promulgator was Svante Arrhenius, one of the greatest chemists of his (or any other) age. Yet while Arrhenius agreed that panspermia was possible, he disagreed with Richter and Kelvin about the motive power. Arrhenius, who had seen enough shooting stars to know that any falling meteorite undergoes a violent heating to temperatures that can melt rock, noted: "The surface of meteorites become incandescent in their flight through the atmosphere, and so any germs which they might possibly have caught would be destroyed." Any casual glance at a clear night sky will sooner or later result in the sighting of a shooting star, the bright celestial light show produced when even small hunks of matter enter the earth's atmosphere. The tragedy of the latest space shuttle disaster shows all too well the enormous temperatures that objects reach when slamming into a thick atmosphere while traveling at more than twenty-five kilometers per second, a typical entry speed for an asteroid or meteor. Comets come in even faster, at speeds of as much as seventy-five kilometers per second. At such velocity most matter turns molten, reaching temperatures far in excess of that necessary to kill life, or so the thinking has gone. So there was certainly reason for skepticism. Arrhenius's view was that life could spread throughout space in the form of tough and impervious spores of some kind and that the motive power of their spread was a force that he called radiation pressure. He saw these spores as escaping the atmosphere of a planet with life and then spreading to other stars by light pressure. But as science learned more about physics, it became increasingly clear that space was a hostile place. In particular, ultraviolet radiation was seen as a potential killer of the spreading spores. But in spite of these doubts the idea took hold, and it has had advocates ever since. Carl Sagan critically reexamined the panspermia hypothesis in the 1960s, and he thought it plausible that some star systems, such as red dwarfs, might catch such interstellar traveling spores. It was yet again sensationally revived in the 1970s by the British cosmologist Fred Hoyle and his colleague Chandra Wickramasinghe, who believed that the process is so widespread and common that aliens are being delivered to Earth each day and that new diseases may be attributed to this. Also at this time, an even more startling idea was proposed by the Nobel laureates Francis Crick and Leslie Orgel, who suggested that life was spread by intelligent beings, leaving microbe-ridden trash on planet after planet or deliberately seeding planets with microbes. This idea came to be known as directed panspermia. Even with the sterling pedigrees of the two sets of proposers, these two rather dubious claims helped put the theory back into the bin of crackpot ideas, but panspermia would not stay discredited for long.

Could life have traveled as the result of light pressure propelling spores through space, or was the motive force for microbial stowaways the passage through space on fast-moving meteors? It is this second idea, now called the ballistic panspermia hypothesis, that seems possible. As far back as the late 1800s the idea of ballistic panspermia was advanced: that it was not light pressure spreading spores, but meteor-carrying stowaway microbes that constituted the best chance for the interstellar transport to life. But what of the pertinent observation that meteors burn when they enter the earth's atmosphere? A quick observation of meteorites in collections at many museums can overturn this objection. While it is true that rocky material from space ending up in the earth's atmosphere heats upon entry, it is only the sand-size fraction that completely vaporizes. Even meteorites as small as a piece of gravel will have an interior that does not melt or even approach a melting temperature. Clearly, the reentry process could preserve microbial life if it were safely tucked into the middle of a gravel-size meteor entering the earth's atmosphere. But any trip through space must begin as well as end. What of the event that ejects a rock off a planetary surface in the first place? Any such event, which could be triggered only by the impact of some large comet or asteroid on a planet's or moon's surface, would be highly energetic (there is probably not enough energy even in the most violent volcanic eruption to eject gravel-size or larger rocks out of the gravity well of a terrestrial planet and into space). It is the blastoff from a planet that seems to be the most difficult feat, and this conclusion led Gene Shoemaker, some thirty years ago, to opine that any impact eject managing to escape Mars gravity would necessarily be heated to such high temperature that any life along for the ride would be cremated. But this view did not go unchallenged. More recently Jay Melosh of the University of Arizona took a very different view; he concluded that rock-size material might be ejected off the Martian surface *without any shock heating at all*! He based his work on calculations of the stresses experienced by meteors as they are thrown into space.

How likely would it be that Martian material would land on the earth? Very, it turns out. Computer modeling has shown that as much as a billion tons of Mars has landed on Earth since the solar system formed, and even in the later, more quiescent stage of the solar system, as many as a dozen fist-size rocks have landed on the earth each million years. Finding a Martian meteorite thus does not seem such a chancy event.

The discovery of the Alan Hills, Antarctica, meteorite and the sensational 1996 announcement that it might hold the fossil remains of life that had come from Mars, brought the idea to the fore once again. The whole journey of this fateful rock from space is itself extraordinary. It was part of the Martian planet for more than four billion years and was blasted off Mars some fifteen million years ago (the figure is based on radiometric age of melt products on the meteorite), and it then floated in space for nearly that long. About fourteen thousand years ago it quit its nomadic space wandering and fell to Earth. The whole enormous controversy around whether or not that fateful Martian meteoroid did indeed contain the remains of life may, more than any other factor, has revived the panspermia hypothesis as well as helped bring about the renaissance in astrobiology that is still in full bloom.

The claim by David McKay and others in NASA in 1996 that a meteorite originating on Mars held evidence of Martian life brought the idea that there may be life on other bodies in the solar system to the fore, following a long period of quiescence. This discovery galvanized the White House, and a direct result was the formation of the NASA Astrobiology Institute. The discovery also brought the concept of panspermia front and center again, after a long time of disbelief by most scientists in the idea that life could pass from cosmic body to cosmic body and live. But to be accepted, three things had to be shown: first, that the event blowing a rock off a planet into space would not sterilize any microbial stowaways; second, that the microbes could withstand the rigors of space and also withstand the time span of the trip; and third, that they could survive the intense temperatures of the reentry into a planet's atmosphere. All three of these events had to be tested if the hypothesis was to be accepted even on a theoretically possible level.

The test of whether life could survive the initial ejection into space was the goal of Ben Weiss and Joe Kirschvink of Cal Tech. Kirschvink had been involved with the Martian meteorite controversy when he concluded that the small strings of beadlike structures found on ALH 84001 could have come only from a biological source; to Kirschvink, such structures have been observed to have been made only by organic synthesis in a group of microbes called magnetic bacteria. Thus he was interested in pursuing further ramifications accruing from this possibility. He and Weiss approached the problem of whether life could have survived takeoff (and landing as well) in a novel way: by looking at the nature of the meteorite's magnetism.

The two men found that the outer crust of the meteorite was strongly magnetic. But only a few millimeters into the meteorite the magnetic field was much weaker, and its pattern of magnetic field differently organized from that of the surface. The interior field was one that was acquired while the rock was part of Mars. The presence of this weak field showed that the interior of the meteorite could not have been heated more than 200°C (and probably much less) in its takeoff from Mars or its entry through the earth's atmosphere, or the magnetic field would have been lost or changed. Further tests showed that the interior of the meteorite could not have been heated by more than 100°C, a temperature that many microbes, especially those in a tough resting stage, can easily survive. Thus it was shown that survival of the takeoff and landing episodes was possible. But what of the long sojourn in space?

How long can microbes last in space?

We very much want to know how long a microbe can stay dormant as it travels through space. Since the travel time from place to place is never direct but will almost always involve some long-term holding patterns because of celestial mechanics, this is a critical factor in understanding the possibility of panspermia. Unfortunately, humans have been interested in this question for only a short time, and since many potential dormant periods must exceed human life spans, direct observation of this trait is not going to produce the answer.

The question thus turned to how long a microbe *could* last in space and remain viable. Fifteen million years is a long time. But then new results from microbiologists studying 250-million-year-old salt deposits changed everything. Such rock deposits, from the Permian period, are known to enclose many fossilized microbes. In a stunning discovery, some of these ancient mummies were brought back to life. With this discovery, long-term microbial dormancy became much less an issue or worry for those advocating panspermia. While only two of fifty-three salt crystals studied yielded viable activity (suggesting that survival on this scale is rare), the very fact that some seem to have survived greatly increases the odds that panspermia, at least in the solar system, might be a viable event. This study is not without controversy, however, with critics suggesting that the microbes are much younger.

But if time was not an issue for sleeping microbes, what about the rigors of unprotected space travel? How could any form of life survive the cold vacuum and hard radiation of the outer space environment? Each would seem to be an effective antibiotic. Yet while it is clear that any form of animal or plant soon dies in the vacuum of space, microbes are much hardier life-forms. Life is tenacious, and none more so than the Bacteria and Archaea of our planet and, one would presume, microbes evolved on other bodies as well.

Just how tenacious life can be was recently demonstrated by a team that left bacterial spores on a satellite orbiting the earth for five years. Bathed in ultraviolet light, exposed to the vacuum and cold of space, chancing the hits from cosmic rays, the microbial cosmonauts were resurrected and brought back to life. So it can be imagined that some microscopic spore or microbial resting stage, deeply dormant and perhaps well buried in a speeding comet or asteroid, could indeed make a journey of perhaps thousands or even millions of years of slow drifting in space and finally come to Earth (or some other planet's surface) in a meteor fall or comet impact.

The spallation hypothesis

Various studies thus show that panspermia might work through actions of large-body impact on a planet or moon, blasting rocks infected with microbial life up into space. But exactly how? Which rocks would be thrown up? This was the subject of a study of impact specialist Jay Melosh, who has proposed the spallation hypothesis as a model of how interplanetary panspermia may come about. It remains our best overview of the many components necessary for the successful transportation of microbes from planet to moon or planet in the solar system and is based on the physics of large-body impact.

When a large asteroid or comet strikes a planet or moon, it excavates a large impact crater whose size and shape depend on the size, composition, and velocity of the impacter, as well as the local rock being struck. The impact event generates shock waves, but some of these cancel each other out in what is known as a spall zone, which is located around the rim of the newly excavated crater. Material in the spall zones remains relatively unshocked and unheated. Let's assume that there are microbes living in the rock or soil material that becomes the spall zone. This material will be blasted up to space, but since spall zone material has escaped the brunt of the shock and heat from impact, the microbes will have a greater chance of survival. There are more tests awaiting them, however, as I outlined earlier in this chapter. Outbound from the planetary surface they have been so rudely (and quickly) blasted off of, they will experience acceleration shock and heating from the friction of passing upward through the atmosphere. Any rock material melted in this way will vitrify (become glass) when hitting the cold of space. At least for Mars, the thin atmosphere of the present day would reduce such heating, although for the Mars of the past, this does not hold true. Once in space, any microbes still alive must survive the dangers of ultraviolet radiation and a possibly long wait until they reach a planetary surface again, yet we now think that this is survivable. If we assume that they do crash onto some planet, rather than into the sun, they will have to survive the entry into a new planetary atmosphere, as well as potential ablation and possible breakup as they pass downward, and then they have to survive the shock of impact onto the planet. It is this last aspect that has been neglected. Melosh's work made the community realize that heat on takeoff and landing is not the only issue; there is a huge pressure and shock increase both leaving and, with a thud, arriving. Another issue was neglected as well. Let's say our microbes make it to an alien planet. Sitting in a rock does nothing for them; like the poor Polynesians arriving on island shores after long and perilous trips across the wide Pacific Ocean, our microbes will be weak and "hungry." Once arrived, any still living thus have to be transferred into a living situation. On earth this might mean contact with water-laden sediment or water itself, while on Mars it would require rapid burial of the microbes if they were to survive.

So what are the odds? There can be between ten million to one billion individual bacteria in each gram of earth soil. While more rocks have landed on earth from Mars than the reverse, there has been substantial communication between the two planets. The most extensive review of the odds of viable panspermia between Mars and Earth, and Earth and Mars (by Charles Breiterman) concluded *that it is probable that viable microbes have gone both ways.* Did they live once they get there? We may find out if we get to Mars.

Interstellar panspermia

The original ideas about panspermia by such luminaries as Lord Kelvin more than a century ago were concerned not so much with transport within our solar system as between stellar systems. Again, Jay Melosh has reexamined this hypothesis and found it wanting.

Using a mathematical simulation involving Monte Carlo statistics, Melosh found that chances are that only one or two bits of solar system rock have made their way into a nearby star system over the past 4.6 billion years. But even after the rock got into the stellar system, there is only one chance in ten thousand that our far-traveled rock then landed on either a habitable planet or moon. These are odds so close to zero that for all intents and purposes, the possibility of interstellar panspermia, from us to the nearest stars, or from them to us, can be dismissed. Just won't happen.

Martian genesis for earth life?

As I have suggested in chapter 4, there are a pair of investigators, Kirschvink and Weiss, who argue that earth life could not have started on Earth. This view is in no small way abetted by a recent calculation by Melosh and others who think that at the present time at least a half ton of Martian rocks fall into the earth's atmosphere *each year*.

If life could survive the ride, is there any evidence that it has done so, traveling from planet to planet in our solar system? Kirschvink and Weiss think so and in a highly controversial paper have suggested that the first earth life did not originate on earth at all. Kirschvink and Weiss think that we may be Martians. If so, I may have to rethink the name that I have given earth life. The Greek name for Mars is Ares; perhaps we are not Terroans, but Areons. Better yet, let us hope that life has independently arisen on Mars; in that case we will need not only that name but the name for a whole new tree of life, but that is getting ahead of ourselves, and we'll come back to possible Martian classification later.

Joe Kirschvink gave an address on this subject, appropriately enough, at the Carl Sagan Lecture at the American Geophysical Union in San Francisco in 2001. His reasoning was as follows. If we go back in time to soon after the formation of the solar system, which planet, Mars or Earth, would have been more suitable for the formation of life? Kirschvink argued that metabolism was a key aspect of where life may have started. Life, according to his and Weiss's view, would have more readily evolved and would have had continents already in place. Life needs to oxidize things to exist. Mars is the place of origin for current earth life. And perhaps even more compelling is their reasoning about a place for borate synthesis of RNA to take place—linked impact craters in a desert—which, on the water-covered earth of four billion years ago may have been impossible, as we recounted in chapter 4.

Amino acids from space

So if microbes could make it, what about nonliving organic material? Was the origin of life on Earth—or any other planet or moon— made possible by organics delivered from space? Or can we eliminate this hypothesis because of heating on entry into the earth's atmosphere? Once again we can use the same reasoning that we invoked for testing the viability of cells traveling through space to look at the viability of amino acids traveling through space and landing on another celestial body.

In a series of dramatic experiments undertaken at the Los Alamos National Laboratory, astrobiologist Jennifer Blank and her colleagues demonstrated that not only microbes but even amino acids could survive the ride on a comet or an asteroid or the impact of one on their heads. They used a technique involving a giant gun, virtually a cannon. A soda can–size shell was loaded into the gun and shot at a target of water laced with amino acids. The great energy liberated upon impact mimicked the energy liberated when a meteor hits the earth. This experiment simulated not only meteor impact but comet impact as well. The researchers found that the amino acids were not destroyed by the impact but survived. To the surprise of the investigators, even after the enormous shock and heat of the impact, the amino acids combined to more complicated forms. Freezing the target, to simulate an impact of an icy comet, increased the survival rate even more.

These experiments suggest that the building blocks of life, or even life itself, have routinely passed from planet to planet, planet to moon, moon to planet, and moon to moon. It opens up the possibility that previous sterile bodies in space can have the stuff of life—or life itself, seeded from space.

A new proposition: Ribosa traveled more readily via panspermia than Terroans

In chapter 2 I proposed that prior to DNA life, or even cellular life, there was a stage consisting of a short RNA genome encapsulated in a virallike protein coating. Now I would like to make a new proposal: Many of the varieties of those Dominion Ribosa life-forms (and I include it and its degenerate parasitic descendant the RNA viruses) could have more easily traveled piggyback on ejecta material thrown back into space after large-body impact than more complicated cells, including Bacteria and Archaea. While we now test for the ability of microbes to withstand the rigors of taking off and landing with a comet or an asteroid, I think that "the simpler the tougher" is probably true, and our new and tough ribosans may have been better able to withstand the ride. Perhaps we can modify the Kirschvink and Weiss idea about life starting in Mars in the following way: Ribosans started on Mars, traveled to earth, and evolved into Terroans here. Or perhaps they did all their assembly here. If true, it simplifies the problems of origin of life scenarios. Tough ribosans could be built in one place, get carried across the surface of the planet they started on in some or various fashions, and then end up in another assembly spot, where they could infect larger protocells. The important point is that they were good travelers.

So here let me wade in once again with a modification of the panspermia hypothesis: *Panspermia occurs more readily or frequently early in a life's history, when a life-form has a minimum genome and is already adapted to extreme environments.* In the case of life on earth, I suggest that ribosans were better travelers or were the stock of microbes that seeded life from Mars.

Life beyond Earth?

The work profiled in this chapter makes clear that the possibility of life's originating on one planet and then moving to another is tenable. This is especially true for the Earth-Mars-Venus triplet, and perhaps for Titan and Triton as well, as we shall see. But now let's look for life beyond Earth.

Chapter 8

Mercury and Venus

Long before we made telescopes and spacecraft, before we thought about orbits and spinning spherical worlds and the unique relationship between Earth, our home world, and Venus, the world next door, people everywhere knew there was something peculiar about Venus.

—David Grinspoon, *Venus Revealed*

Amateur astronomers are well aware of the huge difference between those planets sunward of Earth and those outside its orbit. We are used to the phases of the moon: its waxing from a crescent to full, then its waning back toward a crescent, and finally invisibility. To an amateur astronomer Venus and Mercury have the same life history. When full, our moon can positively glare, but Venus, which also changes noticeably in brightness, is brightest to us when it is a crescent and dimmest when it is full. In contrast are the outer planets, from Mars out toward the gas giants. They never show phases; they always appear full to us. They also show geography that anyone can see. In contrast, Venus and Mercury are but bright white lights, close but far more mysterious than the outer planets. In this chapter we shall look at Mercury and Venus as abodes for life. Is there any hope? Was there ever life there? And what sort of alien microbe might survive in these sunward planets of the solar system?

Could there ever have been life on Mercury?

Mercury is small as planets go, and were it orbiting Jupiter or Saturn, it would not even be the largest moon of these two systems. But just because it is the size of a Jovian or Saturnian moon does not mean that Mercury is anything like those far-distant satellites. Mercury is the densest planet in the solar system; it has an iron core the size of our moon, surrounded by a lower-density crust. Moreover, the Jovian and Saturnian satellites are rich in water; Mercury is rich in metal.

Mercury is obviously a lot closer to the sun than Earth, and as the sun is eleven times brighter than on Earth. With no atmosphere to speak of, the daylight side on Mercury is far hotter (850°F) than any temperature ever gets on Earth, while the night side is far colder (−300°F) than minimal Earth temperatures. All in all, not an earthlike habitat. Although Mercury seems like one of the longest of long shots in the search for life in the solar system, the discovery of what might be substantial volumes of ice at its poles has changed the equations somewhat. Radar imagery has observed steep-walled craters in the polar regions of Mercury. Because Mercury's tilt (axis of rotation) is at such an angle these high-latitude sites never receive direct sunlight.

A NASA spacecraft, successfully launched in August 2004, is heading toward Mercury. The *Messenger* spacecraft will pass by Venus twice and not enter orbit in the Mercury system until 2011 (after two earlier flybys in 2008 and 2009). There is no life detection equipment on this spacecraft, which will not attempt any landings on the planet's surface.

Is there any hope for life on Mercury? Doubtful. No liquids. Unless there can be life in ice, and this is disputed even on our Earth, there seems no hope.

Venus

After the sun and moon, Venus is the brightest object in the sky. But it follows a loopy schedule in its movements. How could the earth's early sky watchers reconcile the fact that Venus (and Mercury, though much less conspicuously) alternate between bright objects following the sun after sundown and preceding it in the early-morning sky before dawn, but never appear on the zenith and never march across the night sky like the other planets and stars? This bizarre pattern was a major impediment among those trying to understand whether Earth was the center of the universe, around which everything revolved, or whether there was no other order to the universe.

Venus at first seems like the *least* likely place in the solar system that could support life—at least now. It is hot. Temperatures at the surface exceed 900°F, and the atmosphere, as documented by a succession of space probes over the last four decades, is a poisonous miasma of carbon dioxide, chlorine, and sulfuric acid. Such temperature and noxious chemistry rule out any sort of carbon-based life on the planet's surface.

That Venus is so hot was a major surprise. In the 1940s and 1950s a succession of science fiction novels using the planet as a backdrop portrayed a warmer but still habitable twin of Earth. Since Venus is about three-quarters of Earth's distance from the sun, it was expected to be hotter. But 450 degrees hotter? The answer came when the thickness of the Venusian atmosphere was first measured. It is a hundred times thicker than Earth's atmosphere, and because it is composed largely of the greenhouse gas carbon dioxide, the Venusian atmosphere traps heat against the planet. That heat comes from two sources: the nearby sun and the intense heat flow emanating from the planet's interior. These two factors make the surface of Venus hotter than any other planet or moon in the solar system.

Venus present is a vision of hell. But what of Venus past? Early in its history Venus might have been the most favorable spot in the solar system for earthlike life. It surely had a thick atmosphere and perhaps liquid. Lakes or even oceans might have been present, and while Venus is closer than Earth is to the sun, early in solar system history the sun was 30 percent less energetic than it is now. Being dimmer would have made the closer Venus a world akin to present-day Earth in terms of the sun's influence on planetary heating. If this view is correct, Venus past might have been stocked with life, perhaps enough to leave a fossil record and perhaps enough to have been blasted off the planet into space, thus potentially infecting other planets with microbes. Earth would have been the most likely target. One of the most astute of all planetary scientists, Kevin Zahnle of NASA Ames, thinks that Venus was once habitable and probably inhabited. Zahnle is a guy with the kind of track record that you bet your money on. The planet named for love may have had plenty of it.

Venus lost all of its earthlike qualities. The planet has undergone a runaway greenhouse effect, losing whatever surface water it may have had to space. In the process of losing its water, Venus evolved a thick and noxious atmosphere that crushes the surface under gas pressure ninety times greater than that on Earth. Furthermore, the planet itself is intensely volcanic, and there is enough heat flow from its deep interior to melt the crust to slag occasionally. So even if there was a fossil record, it has surely been destroyed. This insight came largely from the successful NASA *Magellan* probe, which, using a powerful radar system, during the 1990s orbited and mapped Venus. With the thick cloud cover from the roiling carbon dioxide blanket that Venus calls an atmosphere, any sort of visual inspection has proved futile. But the powerful radar beams bounced off the surface of Venus by *Magellan* yielded some unexpected surprises. The biggest is that the current surface features, and perhaps the planet's surface itself, seem to have formed at roughly the same time, solidifying from a molten magma. This too probably happened on Earth soon after its formation some 4.6 billion years ago. But the surface of Venus is only about 700 million years old, a time when the first animals on Earth were diversifying. No one believes that Venus had a magmatic surface for almost 4 billion years. The heating was recent and spectacularly catastrophic. In the end, the melting of the Venusian surface probably removed all trace of any possible ancient life from the planet's early history. (For a paleontologist like me, a possible fossil record melted is the ultimate frustration.) How did we come to learn this?

The radar record of Venus's surface showed volcanoes, highlands, lowlands, and craters. But the surprise was that the craters were evenly distributed. Unlike our moon, which has heavily cratered older areas and more recent lava flows (the ill-named "seas" of the moon, such as the Sea of Tranquillity), Venus shows far fewer craters. This is because the newer seas on the moon have had less time to accumulate craters, and some that were already there were buried in the huge outflows of lava. This too happened on Venus— except the great lava flows occurred almost simultaneously all over the planet, some seven hundred million years ago.

Why? Our planet maintains the process of plate tectonics. Heat rising from the earth's interior heats the base of the crust sufficiently to make it a fluid, and great upward flows of molten rock (the so-called convection cells) carry the hard crust on its back, causing continents to drift. But plate tectonics can occur only in the presence of large oceans; it takes water to lubricate the subduction zones (the place on Earth where the surface crust dives back underground along long submarine trenches). When Venus lost its oceans, it lost plate tectonics (if it ever had it). It now has what we call stagnant lid tectonics: Heat rises over long periods of time and finally melts the crust, all of it. One of the key points of Brownlee's and my book *Rare Earth* was that plate tectonics might be a requirement of an earth-like planet that has oceans for long periods of time. Did Venus lose its oceans because plate tectonics stopped there? This is my supposition: Venus, a good planet gone bad, through the loss of plate tectonics and hence the loss of its planetary thermostat system. Whatever happened long ago, Venus now goes for long periods of time with little tectonic activity and then goes berserk, melting its rocks all at once. This is what the *Magellan* spacecraft (and the very clever scientists who received the data) discovered and taught us.

All in all, Venus seems to be a sterile place. There is no current microbe on Earth that could conceivably maintain life on the surface of Venus, so it *must* be sterile.

Or is it? In a rather sensational and clever feat of imagination, several planetary scientists, including David Grinspoon, have suggested that one part of Venus, its upper atmosphere, might still be habitable and perhaps inhabited. This line of speculation was first published in Grinspoon's excellent treatment of Venus (*Venus Revealed*) and is reiterated in his follow-up, *Lonely Planets*. Under this scenario, the churning Venusian atmosphere might support tiny microbes in its upper, relatively cool cloud banks. This aerial plankton might tap ultraviolet life as an energy source. Grinspoon has even suggested that there might be lichen atop the highest reaches of the five-mile-tall volcanoes of Venus. His theory (at least about the upper atmosphere; the suggestion about lichen has yet to be echoed by any sober astrobiologists) has been taken up by Dirk Schulze-Makuch and Louis Irwin, who have based their suggestions on actual *data*, of all things.

Schulze-Makuch and Irwin analyzed information gathered by several of the Russian *Venera* space probes of several decades ago. Contrary to expectations for such a toxic and poisonous steam bath of a planet, there is a suggestion of water droplets in the upper atmosphere of Venus, and temperatures there might not exceed 70°F. Schulze-Makuch and Irwin noted too that, also against expectations, there is very little carbon monoxide in the Venusian atmosphere. Yet there should be, for lightning and solar radiation would be expected to produce prodigious volumes of this gas by chemical means. They explain the absence of most carbon monoxide (and the equally curious presence of hydrogen sulfide and sulfur dioxide) as indicative of life—life that is metabolizing and transforming atmospheric chemicals into molecules that should not be present in the absence of life. Even more suspicious for those who suspect that the upper atmosphere of Venus might sustain and harbor life is detection by the Russian probes of carbonyl sulfide, a gas so difficult to produce without the aid of life that some consider it an unambiguous indicator of organic processes. On Earth, at least, the formation of carbonyl sulfide and hydrogen sulfide is most easily accomplished by bacteria. Thus there are two possibilities: Either Venus is inhabited by microbes producing these two chemicals (and more interesting, the microbes would probably be earthlike life in being self-similar in this chemistry), or the very weird and un-earthlike atmospheric composition of Venus has accomplished chemical reactions that do not normally take place on Earth. How to choose? By returning, of course, and that may be soon in the cards. A European Space Agency mission to sample the upper atmosphere of Venus and bring the sample, including any microbe living there back to Earth, is in the planning stages.

Grinspoon goes even farther. He is one of the cadres of planetary scientists who have studied Venus, especially its cloud layer, in detail. He notes that there are complex, stable, and mysterious dark stains above Venus that are global in extent and that contain lighter and darker patches that just conceivably could be caused by photosynthetic organisms floating in the clouds. Could any sort of earth life exist like this? Perhaps, for the recent discovery that some bacteria can exist in very acid conditions on Earth removes the obstacle of high acidity in the Venusian cloud layers as a deal buster. And certainly the temperatures there could be easily withstood by any number of earth microbes. What about an energy source? That too would be available through several pathways, most obviously including light. The real problem is getting life not to live in the clouds but to form in the clouds. Just because life can exist does not mean that it can form. Grinspoon wonders if ancient Venus had life, which had to flee into the sky when the oceans were lost and the planet underwent its catastrophic runaway greenhouse event.

Could Venusian life have made it to Earth?

As we have seen in chapter 7, there are many reasons to believe that bits of comets and moons get to fly on Transsystem Spaceways via impact processes. Could Venusian life have made it to Earth or elsewhere? There is certainly a chance of this early in Venusian history, if Venus ever evolved surface life. But later, when and if Venus got the hypothesized "cloud life" speculated upon by some in this chapter, it would be very unlikely that the cloud Venusians could go anywhere. It takes a hunk of rock to protect a microbe on the Panspermia Express as material is blasted off a planet and then perhaps reenters another.

Cloud life would never have the benefit of the life-preserving rock cover, even if it could, somehow, get ejected out of the upper atmosphere and into space.

A search for life on Venus?

How will we test for life on Venus? There is certainly enthusiasm for more NASA and perhaps ESA missions to Venus. But the fate of all past probes landing on Venus has been a quick death in the great heat and pressure. To search for life in the upper atmosphere of Venus would require a very special kind of probe, and there is nothing on the books at this time that could do the job. My view? Not worth the effort to undertake a mission whose major goal is to search for Venusians. Maybe some millennia down the pike, but let us husband our resources for trips out away from the sun, into the cold where life might be, the subjects of the next chapters.

Chapter 9

Fossils on the Moon

Though most asteroids stay in the Asteroid Belt, there is one group—
called the Apollo asteroids—whose orbits cross the Earth's orbit.

—Amir Aczel, *Probability 1*

In the last chapter I described the ongoing missions to Mercury and Venus and the possibilities of life there. But in all probability, no human will ever set foot on Venus; a sample-seeking return mission seems impossible as well. So where might we find bits of Venus rock, perhaps with fossils in them? The answer might be in front of our faces each month. Of all the bodies in the solar system, nothing seems quite as dead—or as little promising as an abode for life—as our dear old moon. Imagine that we were Jovians, with four large satellites within easy rocket reach, or Saturnians, with murky Titan so close by. Instead we have our cold, dour Luna. Clearly dead, and just as clearly, it has always been dead. So why talk about it in a book about aliens? The answer, the subject of this chapter, might be surprising. The moon is indeed dead, but that very deadness might make it of enormous interest both to those searching for clues to how life may have started on Earth and to those interested in the search for aliens. The moon is a nearby museum, and its surface might be a solar system clearinghouse of fossil life from around our solar system. There may be fossils on the moon, from the early earth, from Venus, from Mars. For that reason alone, we humans have to go back.

An unusual proposal

Throughout this book I have made mention of the NASA Astrobiology Institute (NAI), which has catalyzed and funded so much of the new research discussed in these pages. But in Seattle, at the University of Washington, NAI is but one of the legs of our astrobiology program. One of the goals of the institute is to try to bring together scientists from different disciplines to tackle problems concerning astrobiology, a new discipline that examines the origin, makeup, and extinction of life in the universe. The National Science Foundation has also funded us in its effort to create a new type of education, one based on multidisciplinary approaches: the IGERT program, or integrative graduate education research training. The old paradigm of graduate teaching has been to train students intensively in a narrow discipline. One has depth but no breadth, and this approach has certainly served United States science well. But the deeper thinkers administering American science have realized that while this system has indeed worked wonderfully well, there might be another approach that could further science itself—that is, multidisciplinarity. Keep depth, but add breadth, and in perhaps this fashion new insights might be gained. Easy to say, *really* hard to do.

The field of astrobiology is an exemplar of this approach. In 1998, we at the University of Washington began to admit grad students who have had several supervisors , and one of the first fruits of this labor relates to the moon.

As a problem exercise, John Armstrong, an astronomy grad student, and Lyd Wells, an oceanography grad student, joined with Guillermo Gonzales, a postdoc in astronomy. Their task was to think about a vexing problem: the fact that there are no known sedimentary rocks on Earth older than about 4 billion years. The earth is 4.6 billion years old, and while it is well known that life dates back to 3.6 billion years, most astrobiologists suspect that it may have started earlier. The result of this collaboration was a startling proposal: that there are pristine rocks from early in the earth's history on the moon. These are the rocks necessary to test ideas about the timing of the first life on Earth, rocks that were transported to the moon during the heavy bombardment phase of early earth history, when giant asteroids pounded the planet, causing a large amount of material to be transported from planet to planet.

While the UW team was concerned only about early earth rocks, others saw that an even bigger prize might be found on the moon. During the heavy bombardment period Mars and Venus were also pummeled, raising the possibility that fossils not only from the early earth but from these other two planets may also now reside on the moon.

The main insight leading to this idea came from a realization that the time of 4.5 to 3.8 billion years ago was when the earth (and the rest of the inner solar system as well) were bombarded by asteroid impact. There was much space junk in the form of planetesimals left over from the formation at that time of the solar system, which came into existence about a half billion years before the onset of the heavy bombardment. Many of these asteroid strikes on the earth were so violent that the material blown upward would have reached escape velocity. This certainly happened during the 65-million-year-old Cretaceous impact event (indeed, there could be bits of the Gulf of Mexico up on the moon as a result).

The UW team began by talking over various ideas, and somehow heavy bombardment came up. It is a frustrating fact of life that there are on earth very few pristine earth rocks of that ancient age; the tectonic travails of our planet have a way of eroding, heating, or pressuring rocks, and the older they are, the more chance that such early activity has erased ancient chemical signals or even fossils. So the group, knowing that asteroids can throw rocks into space, as we have seen in chapter 7, began thinking about the moon as a witness plate and planetary attic, a place where the bombardment history would be preserved (since the moon has no climate or tectonics, it keeps much ancient history in good condition). Armstrong, Wells, and Gonzales began to do some back-of-the-envelope calculations and were amazed by their results: As much as twenty *tons* of earth rock—material that made its way to the moon during the time interval of 4.5 to 3.8 billion years ago and rests there still as fragments of the lunar soil—might be found in every hundred square kilometers of the moon's surface.

The trio, now excited by their initial calculations, went to work in earnest. They soon concluded that the earth material would not be uniformly distributed on the moon but would accumulate in specific areas. Most, it appears, would be found on the familiar earth-facing side of the moon.

NASA may have already brought back some of this material from the moon. Of the nearly thousand pounds of moon rock returned by the Apollo program of a long-ago and braver America, geologists studying this treasure trove had already concluded that as much as 2 percent of lunar soil returned from the moon was not of moon origin; it came from asteroids or from planets, Earth, Venus, and Mars being the most likely. Such material is easily distinguished from native moon rock by its chemical composition.

So, how to get this stuff? That would require a trip back to the moon. Just such a mission (or missions) is being contemplated by NASA, the European Space Agency, and the Chinese government. Both unmanned and manned missions are being considered.

Interest in a return trip to the moon by the U.S. space agency was greatly boosted by the January 14, 2004, speech of President George W. Bush, outlining his ambitious goals of a manned mission to Mars, with trips to the moon as part of the engineering package. The emphasis on moon and Mars exploration, with robots first, followed by astronauts, set NASA brass abuzz. Directives were sent out to the various units within the agency to begin planning. The NASA Astrobiology Institute was not excepted.

One of the earliest worries by both NASA and the scientific community was that soon after the Bush announcement, there was no trickle-down request for new science to accompany the ambitious new proposal for missions to the moon and Mars. We all know what a deep and scientific thinker our president is. Soon paranoia began to accumulate along with the normal chatter between scientists. Would this happen? Was it a trial balloon? Was it politically motivated? Was he serious? Somehow, out of this, came a rumor that the new missions would *not* be scientific missions. Science, or so the rumors said, would have a backseat. There were whispers of the United States militarizing these missions—and the moon itself. About the same time, the Chinese successfully sent a man into space and got him back, fueling further speculation on the geopolitical aspects of a new moon mission. The pot was near boiling.

In early 2004 it was therefore decided that a meeting of lunar specialists and other interested observers should come up with a list of science objectives that might be accomplished on a lunar mission. I was asked on short notice to fly to Houston to join in. About thirty specialists from a wide range of fields assembled and, over about three hours, hashed out the main points that the geological, astronomical, and astrobiological communities would like to see if we humans soon did go back to the moon. As in the past, during the Apollo missions to the moon, much of the early discussion centered on the geological explorations. But at this meeting and happily for astrobiology, one of the leaders was Ariel Anbar, yet another member of the UW NAI team, joined by Bruce Jakosky, head of the University of Colorado NAI, and in the crowd were a few other card-carrying astrobiologists, myself included. Astrobiology was thus well represented, and a strong case for new directions of research that a new round of lunar missions could accomplish was crafted.

Months later a NASA Astrobiology Institute white paper appeared and was duly communicated to the politicians who decide where NASA goes next. Whether it ever made it in any way, shape, or form into the corridors of power can only be speculated on. I will reproduce the entirety of its executive summary, for it details why we need to go back to the moon and why it matters to those interested in life on any planet, even if it is of little interest to those born-agains who now rule our planet:

> The Moon preserves unique information about changes in the habitability of the Earth-Moon system. This record has been obscured on the Earth by billions of years of rain, wind, erosion, volcanic eruptions, mountain building, and plate tectonics. In contrast, much (most?) of the lunar surface still contains information that reflects events at the time of life's origin and subsequent evolution on Earth. Therefore, lunar research can address critical astrobiology science questions. In particular, the lunar record allows us to focus on two specific issues in the early Solar System—the history of impacts and the history of exposure to radiation. The Moon, as Earth's closest neighbor, is probably the only body in the Solar System where we can address these issues quantitatively. Impacts probably played an important role in the earliest history of life on Earth. Large impacts would have temporarily altered the environment and created hostile conditions in which life could not survive. Later impacts probably shaped life's evolution by forcing successive mass extinctions of large numbers of species. The terrestrial impact history is better recorded on the Moon than on the Earth. Central science goals are to determine the impact rate onto the Moon (and, by extension, the Earth) during the period when life was originating early in Solar System history, as well as in geologically recent times. We can use the beautifully preserved record on the Moon to help us to understand the habitability of the Earth at the time of life's origin and earliest evolution and determine the frequency of impact-driven mass extinctions and the subsequent course of evolution. During Earth's earliest history, its surface also was bombarded by high-energy particles associated with solar activity (from a solar wind that was enhanced during early history and from solar flares) and galactic cosmic rays, and possibly from nearby supernovae and events associated with gamma ray bursts. This bombardment must have had deleterious effects on life at the Earth's surface, and may have severely affected the formation and earliest evolution of life. These ancient events are recorded in the lunar regolith, formed

throughout lunar history by the impact of micrometeorites and which were buried and preserved by subsequent lava flows. Sampling the effects of this radiation within these fossil regoliths, then, provides a window into the energetic-particle environment at the time that the regolith was buried, and sampling many different locations can provide detailed information over time. This will provide a better understanding of the environmental and evolutionary effects of changes in solar activity, of episodes of harsh radiation, and of energetic particle influx from outside the Solar System. Each of these problems can be addressed in a step-wise manner by a lunar science program that includes orbital imaging and remote sensing, in-situ analysis from landed spacecraft on the lunar surface, robotic sample-return missions, and human-exploration missions.

This report makes clear that we may live on a dangerous hunk of real estate. It also can be read as suggesting that the origin of life and the chemical systems themselves are a product of a tumultuous history. In terms of the geopolitical landscape, the moon will certainly become the new Antarctica, a place that no one covets for its climate but that nevertheless has political importance and thus must be staked out.

The importance of the lunar record—tracking mass extinctions

One of the most interesting and least understood questions about the history of life on Earth is the way that external perturbations affected it. We know from DNA studies of modern animals that many underwent genetic "bottlenecks" caused by near extinction. We also know that animal life on Earth was repeatedly affected by mass extinctions. But we have no idea how mass extinctions may have affected earth microbes or other potentially present forms of life, such as ribosans, protein life, or even clay life. It is the study of mass extinctions that I do for the NASA Astrobiology Institute. The safety of our planet and its potential longevity as a habitable planet for life, based on an understanding of past events on planet Earth, are my particular niche and daytime research job.

Mass extinctions have the potential to end life on any planet where it has arisen. On Earth we know of about fifteen such episodes during the last five hundred million years, five of which eliminated more than half of all species then inhabiting the planet, but we know nothing about the time before animals arose. These events significantly affected the evolutionary history of Earth's complex animals. For example, if the dinosaurs had not been suddenly killed off following a comet collision with the earth sixty-five million years ago, there probably would not have been an Age of Mammals, since the wholesale evolution of mammalian diversity took place only after the dinosaurs had been swept from the scene. While they existed, mammals were held in evolutionary check. Mass extinctions are thus both instigators of and foils to evolution and innovation. Yet much of the research into mass extinctions suggests that their disruptive properties are far more important than their beneficial ones—at least for animal life.

What about microbes? Unlike animals, which are fragile and easily killed, microbes are surely less susceptible to mass extinction events. Once established as a deep microbial biosphere, the bacterial grade of life might be difficult to eradicate from a planet—if such communities can live in the absence of a surface biota. As we shall see, a group of astrobiologists from Cal Tech thinks that a subsurface life cannot exist in the absence of surface life. They have proposed this as a reason for there possibly not being life in the oceans of Europa, and if they are correct, it means that the time of heavy bombardment may have been extremely calamitous for any microbial life on Venus, Earth, or Mars. It may be that life on the surface of all three planets was repeatedly sterilized during the period of heavy bombardment, about four billion years ago. This is why the new proposals to look at the lunar record are so important. The dated crater record on the moon will give us a sense of the impact history. But impacts are not the only agents of mass extinction.

The history of radiation hitting our Earth is also relevant. While no evidence of this may remain on Earth (other than the unexplained disappearance of organisms), the moon may retain a record of energy hitting its surface by any cause. Some of this could be from solar flares, or catastrophic bursts of energy emanating from our sun and sweeping outward into the planets like a lethal tide. High levels of radiation may also come from outside the solar system. Astronomers have detected sudden bursts of intense gamma ray radiation being emitted from various galaxies (gamma rays are the most dangerous radiation emitted by atomic bombs). These short but extremely violent releases of energy would be hugely lethal for any life on nearby planetary systems. Very little is yet known about them.

A new entry into the mass death rogues' gallery is cosmic ray jets (also known as gamma ray explosions), which are lethal bursts of radiation produced by violent stellar collisions. Both gamma ray bursts and cosmic ray jets might result from the same source, merging neutron stars. Astronomers Arnon Dar, Ari Laor, and Nir Shaviv have postulated that cosmic ray jets may account for several of the major mass extinctions and might explain the rapid evolutionary events following mass extinctions. They propose that high-energy fluxes of cosmic rays follow the merger or collapse of neutron stars, themselves the residues of supernovas. These explosions are the most powerful explosions in the universe, releasing as much energy as the entire output of a supernova in a few seconds. When two of these objects coalesce, they create a broad beam of high-energy particles that, if hitting the earth, would be capable of stripping away the ozone layer and bombarding the earth with lethal doses of radiation. Radioactive gases would be produced in the atmosphere and then spread over the planet by atmospheric winds.

Finally, we may detect a record of nearby and perhaps lethal episodes of radiation caused by nearby supernovas, for another potential mechanism producing a mass extinction is the occurrence of a supernova in the sun's galactic neighborhood. Two astronomers from the University of Chicago calculated in 1995 that a star going supernova within ten parsecs (thirty light-years) of our sun would release fluxes of energetic electromagnetic and charged cosmic radiation sufficient to destroy the earth's ozone layer in three hundred years or less. Much recent research on ozone depletion in the present-day atmosphere suggests that removal of the ozone layer would prove calamitous to the biosphere and species residing within. A depleted ozone layer would expose both marine and terrestrial organisms to potentially lethal solar ultraviolet radiation. Photosynthesizing organisms including microbial phytoplankton would be particularly affected.

On the basis of the number of stars within ten parsecs of the sun in the last 530 million years and the rates of supernova explosions among stars, astronomers have concluded that it is *very* plausible that there have been one or more supernova explosions within ten parsecs of the earth during the last 500 million years and that such explosions are likely to occur every 200 to 300 million years.

The frequency of these events is clearly the critical issue. New calculations suggest that these events may be both more frequent and more dangerous to life in any galaxy than previously supposed. Chicago physicist James Annis proposed in 1999 that gamma ray explosions are so lethal that a single such event could potentially sterilize life over much or all of an entire galaxy. Annis has calculated that the rate of such explosions is about one burst every few hundred million years in each galaxy. He suggests that if the energy from such an event hit the earth, it would kill all land life on the planet, even if the explosion had occurred at the center of our galaxy. If such violent and dangerous collisions are rare, they are but one more low-probability event. Yet both Annis and Dar argue that such collisions occur relatively commonly and were even more common earlier in the history of the universe. They also calculated that such effects would affect the earth sufficiently to cause a major mass extinction each hundred million years.

For all these events the moon becomes a witness plate, a real record of the cosmic catastrophes suffered by our planet since its birth. Do we live on a safe planet? How long might our species expect to live in safety before cosmic catastrophe is once again aimed our way? By analyzing the moon's soil, we might answer some very important questions about our future and how much we should worry about things going bump in the starry night.

Back to the moon?

The moon mission envisioned by the U.S. government was meant to be a stepping-stone to Mars. But as I have shown, there are critical gaps in early earth history—when life was first forming—that only the moon's surface might answer. Machines sent up there can't do the job. Only humans can.

The moon always has been dead. But scattered across its dusty surface are fragments from a long-ago time when there were probably three blue planets circling our sun. In the next chapter we will go to the outermost of these and the place where there may be the best chance to find life not of this earth, Mars.

Chapter 10

Mars

It was not by coincidence that scientists anticipated the Mariner-Mars mission by inventing a new science: Exobiology, or the study of life beyond the Earth.

—William Burrows, *This New Ocean*

On January 3, 2004, NASA pulled off one of its greatest triumphs, the landing of a rover on Mars, and then, a week later, it did it again. The rovers, named *Spirit* and *Opportunity*, made their way to the Red Planet, the place that many think has the best chance in the solar system of harboring life. And if life is there, most likely it will be a real alien, the life as we don't know about that is the subject of this book. The twin triumphs are testament to the calm expertise of the Jet Propulsion Laboratory. A quarter century earlier NASA planetary probes accomplished two even more audacious landings. The giant machines *Viking 1* and *Viking 2* dwarfed the tiny rovers now on Mars. Unlike the rovers, which are designed for acquiring primarily geological information, the *Viking*s were sent to Mars for but one purpose: to search for life. Paradoxically, they convinced NASA and almost all scientists studying the data that Mars is barren and set back the field of astrobiology by decades. The story behind those missions is relevant to the current

Martian landscape captured by rover images, 2004. (Courtesy NASA)

research agenda of those studying Mars and speaks volumes to what people thought about aliens and what they would be like. NASA has had huge triumphs in this new century on its Mars missions and knows more than it is telling about the possibility of life on Mars. For NASA, information about life on Mars is the crown jewel of its intelligence, and getting that information is a bit like getting the CIA to divulge what it really knew about Saddam Hussein's WMD, even for us insiders. Needless to say, however, there have been break-throughs. In this chapter we take a long look at the Red Planet, home to so many supposed inimical aliens, some of them imagined by the scientists of prior Mars.

The *Viking* disappointments

In 1975 the first *Viking* lander sat amid barren sand dunes, its camera panning the dreary landscape, sending images to a far-off television monitor. On a nearby sandflat, perhaps a hundred yards from the lander, four images could be seen moving in fashions that were decidedly indicative of life, and complexly behavioral life at that. One moved in bursts, followed by periods of stillness. A second moved across the sand in a sinuous, serpentine glide. The two others were more sessile, more like lumps of rock than anything else, but an unmistakable moving appendage or two also argued for the presence of life. These four images were proof of that most valuable commodity—life, unmistakable life, and the rarest grade of life at that. Not microbes, or slime life of any sort, or even complex plants, but animal life, large and vibrant. The *Viking* had shown that it could detect life. Unfortunately, this particular *Viking* was a test, and it sat on earth, not on the surface of Mars.

The lead scientist of the project, Carl Sagan, had to be pleased, for the camera had been fitted to the *Viking* landers at his insistence, and the test showed that it could detect any large Martian animals that were so stupid as to wander into view. Granted, the billion-dollar-plus *Viking* lander was designed to look for life on Mars. This was the flagship mission of the field called exobiology, the search for life beyond the earth, and if all went well, its camera would also send back the first pictures of the Martian landscape, which would be of enormous interest to a second group of NASA scientists, those interested not in life but in planetary geology. Yet most scientists in both camps (with the obvious exception of Sagan) agreed that any Mars life would be microbial, and the detection of microbial life could not be carried out photographically. Sagan had a different and more exalted vision of Martian life and insisted that the giant space probe be equipped with a camera that could see large animals moving in real time, for he had told reporters that Mars might actually harbor polar bear equivalents. A real-time camera placed on the lander was only part of Sagan's desire for the *Viking* instrument. He wanted far more. Just before the launch he said: "I keep having a recurring fantasy that we'll wake up some morning and see on the photographs footprints all around *Viking* that were made during the night but we'll never get to see the creature that made them, because it's nocturnal." Sagan's solution was that the *Viking*s be equipped with large lights to illuminate the surrounding areas at night, and that the landers should be equipped as well with bait, so as to lure the larger animals into camera range. Each ounce of instrumentation is precious on any space probe, and the *Viking*s were no exception. In many cases even small and sophisticated instruments had to be left off them because of weight limitations. But Sagan was advocating tens of *pounds* of equipment for his search for Martian big game, and he won this fight. The cameras were added, and to test them, he had rented a snake, a chameleon, and two tortoises from a local pet shop to place in front of a *Viking* lander on its 1975 field tests conducted on the windswept landscape of the Great Sand Dunes National Monument. This full-scale rehearsal of the upcoming landings of two other *Viking* landers, at that time well on their way to Mars, came off without a hitch. One can imagine the reception that would ensue if today, in mission planning for a trip to Mars, or Europa, or Titan, any astrobiologist insisted on instruments to detect big game. Our view of what aliens might be like in the solar system has certainly changed in a quarter century.

That there was life on Mars was a public perception held for almost a century prior to the *Viking* missions. In 1877, Giovanni Schiaparelli produced the first "modern" map of Mars, on which he showed a system of what he called *canali*. Although *canali* in Italian means "channels," without the implication of being an artificial feature, the word was commonly translated into English as "canal." In 1910 the American astronomer (and socialite) Percival Lowell further captured the imagination of the public with his book *Mars as the Abode of Life*. On the basis of his extensive visual observations, Lowell painted a portrait of a dying planet whose inhabitants had constructed a vast irrigation system to distribute water from the polar regions to the population centers nearer the equator, thereby explaining the vast system of "canals" that Schiaparelli had supposedly seen. While Lowell's ideas were gradually discarded during the middle twentieth century, his inspired idea of an earthlike Mars proved more durable. At the dawn of the space age, Mars was considered to have an atmosphere about a tenth the density of Earth's, water-ice polar caps that waxed and waned with the seasons, and an annual "wave of darkening" that was often interpreted as growing plant life.

In the 1960s observations from Earth and flyby spacecraft signaled the beginning of the end for Lowell's vision of Mars as an earthlike planet. The *Mariner 4, 6,* and *7* missions returned images of a moonlike, heavily cratered surface. The atmosphere was found to be almost pure carbon dioxide (CO_2), only a hundredth the density of Earth's, and the polar caps proved to be almost entirely frozen CO_2. The first global views of Mars, returned by the *Mariner 9* orbiter in 1972, revealed that the planet was far more complex than the earlier flyby missions had shown, with an enormous canyon system and the largest volcanoes in the solar system. The most interesting observation was of unmistakable flow channels, sure evidence of running water at some point in the past. But the wave of darkening was shown to be the result of seasonal redistribution of windblown dust on the surface, and most of the evidence for an earthlike Mars was swept away.

Despite all these disappointments, the possibility of organisms on or *in* the Martian surface could not yet be ruled out. For this reason, the *Viking* landers carried sophisticated instruments to look for possible life forms on the Martian surface or buried in its soil. *Viking 1* was launched on August 20, 1975, and arrived at Mars on June 19, 1976. The first month of orbit was devoted to imaging the surface to find an appropriate landing site for the *Viking* lander. On July 20, 1976, *Viking 1* finally fired its retrorockets, successfully deployed its parachute, safely landed on rocky terrain west of the Chryse Basin on Mars, and began a photographic survey of its neighborhood.

Its first pictures showed an arid reddish landscape and pink-tinged sky. No large beasts (or even footprints, although Sagan avidly searched for them) were to be seen. But however disappointing this first result was to Sagan and his exobiology colleagues, all the investigators and engineers feverishly readied for the main event of the *Viking* mission, and most regarded the upcoming experiments with an optimism that ranged from guarded to unbridled. A week after landing, after having transmitted a large number of spectacular pictures of the geology surrounding its landing site, *Viking* began its search for life on Mars in earnest.

Viking had been conceived as a multi-investigative program. But while its study of the chemistry and geology of the soil and atmosphere was important, its primary mission, and most of the instrumentation crammed into the crowded spacecraft, was dedicated to the search for extraterrestrial life. This great emphasis on biological rather than physical science components on the *Viking* spacecraft was due in large part to Carl Sagan, and great credit should be given to him for this vision. Sagan had dominated the years of planning and construction, and his belief and optimism that Mars, as well as many other planetary sites beyond Earth, might harbor life were one of the driving forces in convincing NASA, and ultimately the U.S. government, to spend more than a billion dollars on the ambitious attempt to land two large instruments on Mars in the mid-1970s.

By the time that the landers arrived at Mars, NASA, and indeed the whole country, were in dire need of some good news. Watergate had rocked the country and toppled the president, and this on the heels of the agonizing defeat and enormous loss of life in Vietnam that had left the country questioning its compass and fortitude. NASA itself was now several years past its triumphs on the moon, and its next great manned space mission, the first flight of the space shuttle, was still years in the future. Many in NASA hoped that a successful landing of even *one* of the landers on Mars would jump-start the program and lead to what was the next logical step following the Apollo program, a manned mission to Mars in the 1980s. Many hoped as well that a successful discovery of life on Mars would transform NASA from an organization involved mainly in manned spaceflight to one that had a larger scientific component. NASA itself realized that it badly needed a boost. *Viking* was seen as this new beginning, and the fact that both landers had successfully withstood the rigors of launch and the long journey to Mars was itself an enormous success. The landing of the first *Viking* and the successful first pictures seemed to augur even greater things to come.

On July 28, 1976, a robotic claw extended from the spacecraft and scooped Martian soil into the spacecraft. Four basic experiments, all designed to look for chemical evidence of life or its processes, were performed. The initial experiments raised hopes that Mars indeed harbored extant life in its soil, for it was soon found that the soil contained more oxygen than was expected and that chemical activity of the soil at least *hinted* at a microbial presence in the Martian soil. These first-blush experiments created such a wave of optimism in the *Viking* scientific team that Sagan was stimulated once again to raise the specters of Martian big game, telling the *New York Times:* "The possibility of life, even large forms of life, is by no means out of the question." But soon after, a huge dose of cold water was applied to the biologists and their early optimism: The onboard spectrograph, after carefully analyzing the Martian soil, could find no evidence of organic chemicals in the soil. Life, at least life as we know it, leaves behind this inherent signature. If *Viking* landed on virtually any area on the earth, even places like deserts that are seemingly sterile, a similar experiment would reveal the presence of organic molecules. But Mars, at least at this first landing site, contained none. In fact the soil seemed to contain chemicals that would soon destroy any life that might be there. Mars, as viewed from this first *Viking* lander, seemed not just dead but inimical to life. Not only was life not there, but any life that might be there would soon be killed by the toxic chemicals in the soil. Sagan, ever the optimist, could now only hope that the second lander, then orbiting Mars, would be the one to yield the telltale evidence of life.

Viking 2 was launched on September 9, 1975, and entered Mars orbit on August 7, 1976. On September 3, 1976, the second lander safely parachuted onto the Martian surface at a place named Utopia Planitia. Like the first, this huge machine functioned perfectly. Also like the first, no evidence of life was found in any of the separate, and crucial, life detection experiments. Not only was the hoped-for animal-rich Mars of Edgar Rice Burroughs and Carl Sagan nonexistent, but the Mars of a far larger contingent of biologists who speculated that microbial life might flourish in the Martian soil was also apparently a false dream. Mars seemed sterile, and now the main hope lay among paleontologists, who still dreamed of a fossil record telling of the once-living glories of a now-dead planet. Yet fossils were a poor substitute for life, and only theorized fossils an even poorer reason to spend billions on a manned space expedition to Mars.

The *Viking* life detection experiments

Three separate experiments were attempted to look for life. In the first, a nutrient-rich solution (at least a solution that would be delicious for earth life) was mixed with Martian soil samples (which had been mixed with water prior to being fed this nutrient broth) to see if any gas that might be emitted by life could be detected. The resulting concoction emitted a large volume of oxygen and lesser amounts of carbon dioxide and nitrogen. But repeated experiments, while showing vigorous chemical reactions, gave no signs that life was present—earth life anyway. Heating the Martian soil to 145°C changed the reactions, but there was no indication that this happened because Martian microbes had been sterilized. All that could be concluded was that adding water to Martian soil caused a vigorous chemical reaction. Life was not ruled out, but it wasn't detected either.

The second experiment was called the labeled release experiment. Again, a broth was added to a sample of the soil (more accurately named regolith) , but the broth contained radioactive marker chemicals. If there were Martian microbes in the soil, they would take up the radioactive carbon and give it off as a waste product of some type of radioactive gas. The experiment worked. A radioactive gas was detected. When the soil was then sterilized, the gas was not detected, seemingly a positive result showing life to be present. But because of the strange chemical makeup of the soil, there remained the possibility that the release was from inorganic rather than organic pathways. The instrument builders had not foreseen the chemical nature of the Martian soil and were not prescient enough to have put on instruments that might really have detected life—Terroan life or at least its equivalents. Sadly, we have still not put such instruments onto Mars.

The final experiment exposed soil to light and radioactive carbon, again to see if any microbes present would take up the carbon. A positive result was achieved, but it again remained ambiguous because of the strange soil chemistry.

All in all, the three experiments seemed to suggest that life on Mars could not be ruled out but could not be proved either. But the nail in the coffin of life on Mars, at least as far as the *Viking* team scientists were concerned, was the finding that absolutely no organic material could be found in the regolith. Even without life on Mars, some organics should have landed on the surface from comets and asteroids. The harsh ultraviolet radiation seems to have cooked out any trace, and no replenishment—as would have happened if life had been present—had occurred. The scientists explained the positives as false positive for life. The one aspect then not appreciated is that all these tests were made with the assumption that Martian life would be like earth life, of a similar chemistry. This may be why life was not detected; the tests missed an alien chemistry. Other possibilities are that the onboard instruments were not sensitive enough to detect life and that life is present but deeper in the soil or subsurface, to protect itself from the harsh UV of the surface.

There is still another point of view: that the *Viking*s found life three decades ago, and this still remains the contention of Dr. Gilbert Levin, part of the *Viking* life detection team. He not only maintained that there was a detection of life in the labeled release experiment results but thought that he could actually see Martian life on photographs taken by the landers, one of which photographed rocks with some strange greenish stains on them that resemble lichens on Earth. Levin was a loud dissenting voice to the eventual NASA party line that no signs of life were detected, and to this day if one brings up his name in front of NASA mucky-mucks, the room will be filled with a series of poker faces and sky-high eyebrows. He may have the last laugh, however, as indirect evidence of biology on Mars continues to accumulate.

In the aftermath of the two *Viking* missions a huge change occurred in the direction for NASA and its planetary missions. No longer was the search for life the paramount driving force behind NASA (my friend and colleague Don Brownlee has pointed out that the hope for finding life was a driving force behind even the lunar missions). The post-*Viking* era in NASA was one in which unmanned missions were designed for the study of planetary geology, not biology, while the manned space program turned toward the low earth orbit missions of the new space shuttle, rather than the promised next step after the moon, which would have been a manned mission to Mars. Yet even as NASA backed away from what has come to be known as astrobiology, the aftermath of *Viking* coincided with the rise of a whole new generation of filmmakers who began to make hugely popular science fiction movies that portrayed a galaxy teeming with life, a view most famously championed by the *Star Wars* trilogy of that era. The great disconnect that occurred continues to the present day. To the public, at least to the moviegoing public, it was as if life indeed *had* been found on Mars, and even today, in these first years of the new century, it is veritably impossible to turn on a television without finding some alien race joining or fighting humans as they all ply between the stars in their physically impossible, noisy spaceships. But it is only now that NASA is again driven by the search for life beyond Earth, as it was from its start and until the *Viking* disappointments.

The Martian meteorite (again)

As we saw in chapter 7, the interest in life beyond the earth was dramatically intensified following the discovery of the now-famous Martian meteorite known as ALH 84001. This hunk of rock was discovered in the Alan Hills region of Antarctica on December 27, 1984, and then filed away, accidentally misclassified as a more common meteorite type, and forgotten for a decade. It was eventually reexamined and on the basis of its chemical composition was determined to be from Mars. A team of NASA scientists then began to probe it, culminating in the stunning announcement on August 7, 1996, that this particular piece of rock might hold the fossil bodies of Martian microbes in its stony grasp. At the risk of some repetition, let us again consider this contentious hunk of rock with specific questions about how it relates to specific kinds of life.

There were four independent lines of evidence suggesting that the meteorite came from a planet with life. First, the meteorite contained small globules of calcium carbonate, or limestone. On Earth, limestone is most commonly produced with the aid of life or by life itself as skeletal elements, although inorganic precipitation can take place. Second, there were organic compounds in the meteorite, called polycyclic aromatic hydrocarbons (PAHs), which, again on earth, are normally created by bacteria. But are these contaminants from the Antarctic resting site of the meteorite? Third, the limestone blebs contained assemblages of iron and iron sulfide minerals that are out of equilibrium with each other but can be formed together by bacterial activity. Later it was claimed that there were chains of magnetite in the rock. Finally, there were microbial-size structures that could have been fossilized microbes.

Of the various lines of evidence used by NASA scientists to arrive at their startling conclusion, the most fascinating were small rocklike shapes in the meteorite identified as fossil bacteria. And why not? Conditions on the Martian surface today are highly inimical to life: subject to harsh ultraviolet radiation, lack of water, numbing cold. The Mars *Pathfinder* expedition seemed only to confirm this—even for the highly tolerant extremophilic microbes. But what of the Martian *subsurface*? Perhaps life still exists in the subterranean regions of Mars, where hot hydrothermal liquid associated with volcanic centers and possibly small oases maintain a Martian equivalent of Earth's deep biosphere—replete with archaeans?

And if life happens now to be totally extinct on Mars, what of its past? Since the *Viking* landing of 1976, scientists have known that long ago Mars had a much thicker atmosphere and water on its surface. Three billion years ago Mars would have been warmer because of its cloaking atmosphere. Such conditions still would have been too harsh for animal life, but to judge from what we now know about the extremophiles on earth, the early Martian environment would have been conducive to microbes. The extremophiles need water, nutrients, and a source of energy, all of which would have been present on Mars. It may be that life is completely extinct on Mars today. Yet there may be a great deal to learn about ancient Mars in its fossil record, one perhaps populated by the Martian analogs to Earth's extremophiles. Andrew Knoll of Harvard University has pointed out that the Martian fossil record may be more complete for very old rocks than it is on Earth, since there has been little erosion or tectonic activity on Mars to erase the billions-year-old fossil records. He even points out where on Mars to search for fossils: an ancient volcano named Apollinaris Patera, whose summit shows whitish patches interpreted to be the minerals formed by escaping gases, and a place called Dao Vallis, a channel deposit on the flank of another ancient volcano that may have been the site of hot water flowing out from a hydrothermal system within the Martian interior. Mineral deposits (and their isotopic values) there might yield a better sense of whether this system might be analogous to those on earth thought to have been involved in the evolution of Terroan life. Could this be an oven for an earth life recipe?

The environment of Mars

In 2003 we earthlings were treated to a spectacular and long-running show. During that summer Mars came into opposition with Earth, and it was closer than it had been, or would be, for centuries. The bright red beacon shining in clear night skies reminded us why the planet has generated so much fascination, for its piercing roseate brightness cut through city fog and streetlight pollution. Even the big cities were bathed in its ocher light. With just a small telescope the disk showed features: the white polar cap, the darker patches on the reddish surface. Larger telescopes gave a more intriguing view: huge volcanoes, great canyons, and seemingly numberless impact craters covering the surface.

The probes that have gotten to Mars show the surface and how stark it is. Even the most desolate deserts on Earth have a scattered plant or two, and evidence that rain has fallen recently in the form of channels. On Mars there is reddish soil, strewn with boulders. The sky is a pale and washed-out dusk even at noon; the sun smaller, cold, and distant.

For life, at least our kind of life, Mars obviously presents major challenges. The temperature never gets above the freezing point of water on the Martian surface and can dip as low as −140°C during winter and high-latitude extremes. The atmosphere, such as it is, is composed a thin veneer of carbon dioxide, and at sea level (if there were still seas) the pressure is equivalent to an altitude of over a hundred thousand feet, more than four times the altitude of Mount Everest. Because of the thinness of the atmosphere, the surface is bombarded with ultraviolet rays from the sun, even distant as it is. There is also a great deal of radiation on the surface, so much so that it will limit any human stay on the planet. Any astronauts who go had better be past reproductive age or better not want children after returning from Mars.

The rovers and the *Viking*s have had a look and more than a prod at the Martian soil, if soil is what you could call it. On our Earth, soil is composed of a large measure of organic material, the humus that is made up of rotting vegetative material. There is no such organic material in the Martian surface layer, more appropriately called regolith than soil. From a chemical point of view, this layer is corrosive and will oxidize material trying to grow in it. The dryness of the atmosphere creates no brake against dust storms, which are ferocious and can rise up to fifty kilometers into the Martian sky.

The small size of Mars is the main problem. Planet mass is all-important in maintaining an atmosphere, and without this mass what atmosphere was present early in planetary history slowly but inexorably leaks out into space, lost forever. The low pressure inhibits liquid water from pooling at the surface. All in all, a nasty place, at least now. But the past might have been a very different story.

In January 2004, after scares from the software glitch in the *Spirit* rover, its twin, *Opportunity,* chugged up to an outcrop of rock and did a little drilling and some tourist photography. Finding the rocky outcropping was sensational enough. Just as learning about the deep past of the earth requires the sampling and study of sedimentary rocks, so too on Mars there has been a long search for landing sites that would show old sedimentary rocks, rather than soil cover or lava. And find such an outcrop *Opportunity* did, with spectacular results, for minerals in the rocky outcrop have to have formed in water. Here was proof of what planetary scientists had long suspected: Long ago Mars had water—lots of the stuff. But how wet was ancient Mars?

How wet was the subject of new discoveries by the rover *Opportunity* in Eagle Crater early in 2004 and then extended in the summer of 2004 in Endurance Crater. In both places there was indisputable proof of water—lots of water, either a large lake or a shallow sea. This "ground truth" was supported by views from space as well.

Two spacecraft orbiting Mars combined, in the summer of 2004, to give a new estimate of the extent of water that may have been present deep in the past on Mars. As of late 2004, the *Opportunity* rover was parked in what 3.7 billion years ago was the middle of a large lake or small sea. Based on imagery from space and a thermal imaging system that helps show the nature of rock outcrops on the surface of the planet, the size of this body of water was on the scale of either the Baltic Sea or the Great Lakes of North America. This was no little pond, but a body of water at least 127,000 square miles in size. The outcrops being mapped are rich in the iron oxide mineral hematite as well as sulfate minerals that are sure indicators of deposition in water. From the scale of the deposits, it is now estimated that nearly 2,000 feet of strata were deposited on this seafloor. That too is significant, since that much strata take a long time to accumulate, and time, in water, is something that life needs if it is to form and then exist long enough to evolve into anything more complex than the simplest microbe.

Thus we see strong evidence that Mars was rich in water for long periods of time on a planet that had an atmosphere thick enough and rich enough in greenhouse gases, such as carbon dioxide, and perhaps methane, to be warm—perhaps as warm as the earth is now.

This discovery confirmed something long suspected. Photographic analysis of the Martian surface had shown significant evidence of past water movement, even floods. The erosive power of water, moving water, had carved characteristic channel deposits, canyons, and now dry watercourses. Almost all this water is now gone and may have been lost to space billions of years ago. But sometime, deep in that past, there was a Mars very different from now, perhaps wet and warm, and thus a place, like the early earth, that seemingly could have spawned life.

Methane and ammonia

It is not just NASA and its dedicated robots that are finding evidence suggestive of life on Mars. In July 2004 the European Space Agency announced that the Mars Express craft, orbiting the Red Planet, had detected the presence of ammonia on Mars. Its presence in the atmosphere, even at very low concentrations, might be a signal that life is metabolizing on or in the Martian soil, releasing ammonia as a by-product. The discovery was made by an onboard spectrograph, which can detect the presence of even minute amounts of such organic chemicals as ammonia and methane. The curious aspect of the discovery is that there should not be any ammonia present at all, for under the constant glare of unfiltered ultraviolet light and other harsh radiation hitting the planet, any ammonia would soon decompose into simpler chemical constituents. Thus any ammonia in the Martian atmosphere is new and is therefore being replenished. This could be done by two pathways, one of which is the presence of life on Mars, the other by Martian volcanic activity. We know that Mars has has very little active volcanic activity—*if any at all.* Compared with Earth, Mars is quite cold and nearly dead in terms of its heat flow and volcanism. Some planetary scientists think that there are absolutely no volcanic emissions coming from the subsurface. In that case the only source of the ammonia would be life in the Martian subsurface. To date we have never witnessed any active flows of lava on the surface or any eruptions from the volcanoes scattered across the Martian surface.

In a similar fashion, another gas possibly indicating that there might be life on Mars has been found: methane. This has now been detected by a variety of methods. I first heard about it as a rumor from Mike Mumma, like me a principal investigator of the NASA Astrobiology Institute (Mike runs the Goddard node). Soon after, Don Brownlee came into my office, grinning from ear to ear, about the discovery. Don, a very careful soul, thought that the odds were that on the basis of this discovery, life had been detected.

When it comes to such matters, and with Mars in particular, NASA keeps its cards very close to the vest. But the evidence is tantalizing. Methane, like ammonia, can be a by-product of metabolism (think of all the methane produced by the millions of cows on earth, each with a troop of microbes in her complex stomach, emitting methane through the digestive process). Like ammonia, methane does not hang around for long in a planet's atmosphere. But also like ammonia, methane can be a product of volcanism. So for both compounds the question becomes: Has there been recent volcanism going on?

As we noted before, the volcanoes on Mars say no. It is estimated that the last major eruption of the giant volcano Olympus Mons was more than 3.8 billion years ago. In a sense, either way is a plus for the search for life on Mars: Either it is there, giving off ammonia or methane, or both, or it is still tectonically active, a huge boon for those hoping that there are still hydrothermal systems active on the planet, places where life could originate and thrive on an otherwise cold planet. We do know that some of the Martian meteorites found on Earth are composed of volcanic rock only 180 million years old. So, at least while giant Jurassic dinosaurs ruled the earth, there were active volcanoes on Mars. We do not know if there has been more recent activity.

The methane story became even more compelling in late September 2004, when new analyses from the ESA Mars Express orbiter showed that the faint concentrations of water vapor and methane found in the Martian atmosphere overlap. The greatest concentrations of the two gases were found in equatorial regions on the Red Planet. These new data suggest that the source of methane and water is the same and is not uniformly distributed on the Martian surface (or the subsurface, which is where the gas is originating from). The specific regions giving rise to the faint gas traces detected by the satellite orbiting overhead are just those places where NASA suspects the presence of a thin subsurface ice layer beneath the surface rubble that passes for soil on Mars. The layer would be not be unlike permafrost. This would give rise to the faint amount of water vapor detected. But what about the methane? Perhaps there are broad-scale, low-level volcanic processes occurring under this ice layer, and if so, there could be a thin water layer caused by heat flow from below. However, there is another explanation. If there were a small population of methanogenic microbes—those that take hydrogen and carbon dioxide and metabolize them into life—there would be just this signal, at these concentrations. As is often stated, extraordinary claims require extraordinary proofs. No one is yet shouting "life," but in the corridors, over the water cooler, when it is Miller time at the end of the day, the astrobiologists gathering these data are quietly smiling, believing that they have detected life.

What kind of earth life would do best on Mars?

On the basis of these discoveries of 2004, there is a much better understanding of what kinds of environments conducive to life might exist on Mars. The best fit for an earth organism, at least according to rover team scientist Benton Clark, is a sulfate-reducing microbe, such as the genus *Desulfovibrio* found here on Earth.

It does not use light for its energy source; it is not a photoautotroph, one of the main categories of bacterial metabolism. Instead it takes advantage of energy found in rocks and chemicals; it is in the group called chemoautotrophs. Its group, of which there are many on Earth, lives in low- or zero-oxygen environments where there are plenty of sulfates around. By combining free hydrogen with sulfate compounds, the microbes harvest energy and use it to build their organic molecules. Their source of carbon is carbon dioxide, which is plentiful enough in the Martian atmosphere to do the trick. Because the bedrock outcrops discovered in Eagle Crater are composed of sulfate-rich sedimentary rocks, it could be that the rover drilling holes in these outcrops has already polished off millions of Martians with its abrasive unit—if there was water present. But there is not, and that is the problem. It is not too cold, there is food, but there is no source of the hydrogen necessary for metabolism. In the past, however, when there was abundant water present, this obstacle would not have been present, for sunlight would break down water frequently enough to provide the necessary hydrogen.

There is also another earth microbe that might welcome Mars as home. In his extensive review on the odds of panspermia, Charles Breiterman has postulated that the bacterium *Acidithiobacillus ferrooxidans* would flourish on present-day Mars. Breiterman based this conclusion on the fact that *A. ferrooxidans* is capable of anaerobic respiration (a good thing for any Martian visitor, since there is very little free oxygen in the atmosphere, only the amount produced by photo disassociation of the Martian soil), and because of this ability, this microbe can derive electrons from elemental sulfur (which, thanks to the rovers *Spirit* and *Opportunity,* we now know to be in abundant supply) and uses ferric iron as its electron receptor. While surface sediments might be a bit too cool for its taste (*A. ferrooxidans* likes about the same temperatures that humans do), it would do nicely down in the Martian crust, where temperatures will be warmer. So, there are at least two of our bugs that would like Mars, and there are surely many more. This in fact might be a problem because it is clear that microbes stow away on spacecraft, and earth life might be happily growing on Mars, courtesy of NASA and the crashed Mars polar lander as well as the Europeans' *Beagle* lander at this moment.

But what about the cold? Surely this will inhibit the growth of life on Mars. Not so, according to the newest work. Here is a quote from microbiologist (and physicist) Buford Price about the limits imposed by cold on life:

> Our results disprove the view that the lowest temperature at which life is possible is ≈−17°C in an aqueous environment, as well as the remark that "the lowest temperature at which terrestrial and presumably Martian life can function is probably near −20°C." Our data show no evidence of a threshold or cutoff in metabolic rate at temperatures down to −40°C. A cell resists freezing, due to the "structured" water in its cytoplasm. Ionic impurities prevent freezing of veins in ice and thin films in permafrost and permit transport of nutrient to and products from microbes. The absence of a threshold temperature for metabolism should encourage those interested in searches for life on cold extraterrestrial bodies such as Mars and Europa.

While these comments apply to the search for life on Mars, they hold as well for other cold bodies, such as Europa and Titan. Life is tenacious, even in the cold. The craters being investigated may not be home to life now. But there is water on Mars, in the form of ice on the polar caps, and there is enough climatic fluctuation (Mars actually wobbles on its axis far more than the earth does) that the ice periodically moves around, if it does not actually melt. During these intervals all the necessities for life might be present. If Martian microbes used a tough spore stage, as many earth microbes do, we could have a life cycle in which, for perhaps millions of years, the tiny seeds of life lay dormant until awakened by water.

Complex Martian life?

The fossil record tells us that animal life did not arise on Earth until less than a billion years ago, whereas life on the planet predates the first animals by about three billion years. One of the most puzzling aspects of life's history on Earth is this singular gap between first life and first animal life. Many factors were surely involved, but there is irrefutable evidence that oxygen is a necessary ingredient for animal life (at least on Earth), and much evidence shows that sufficient concentrations of oxygen were not present in the oceans and atmosphere until less than two billion years ago. Many scientists suspect that the long time necessary for the earth to acquire an oxygen atmosphere accounts for some or all of the long delay between the origin of life and the origin of animal life on Earth. But is the earth history of a long delay between the first formation of life and the first formation of animals or their equivalents the only possibility? Planetary scientists H. Hartman and Chris McKay make the novel suggestion that this long delay was partly due to plate tectonics on earth and then apply this idea to a theoretical history of Martian life.

It is universally agreed that the rise of oxygen on earth took place because of plant life causing the release of free oxygen, as a by-product of photosynthesis. The most primitive photosynthetic organisms used enzymatic pathways called Photosystem I and Photosystem II; however, these systems operating alone do not release free oxygen. In a later event these two systems somehow were combined nearly intact into a single organism, the ancestor of the cyanobacteria, and Photosystem II was able to evolve to the point where it was able to release free O_2. The latter system may not have evolved until 2.7 to 3 billion years ago. Eventually photosynthesizing organisms such as photosynthetic bacteria and single-celled plants floating in the early seas would have released vast volumes of oxygen. There was probably some small source of inorganically produced free oxygen on the early earth as well. It may be, for example, that ultraviolet rays hitting water vapor in the upper atmosphere created free oxygen, at least in small volumes. However, a net accumulation could not take place until various reducing compounds (which bind the newly released oxygen and therefore keep it from accumulating as a dissolved gas in the oceans or as a gas in the atmosphere) were used up. The amount of iron in the crust of a planet will greatly affect rates, for all of it on the surface in contact with the atmosphere must be oxidized before free oxygen can accumulate. Such chemically reducing compounds (ones that will react with oxygen) emanate from volcanoes, and it can be argued that planets with a higher rate of volcanic activity will have more reducing compounds in their oceans and atmospheres. Another important source of reducing compounds are organic compounds, produced either through the death and rotting of organisms or through the inorganic production of organic compounds, such as amino acids. Great volumes of such material are found in the oceans on Earth, but it is usually buried in sediments. In the absence of plate tectonics, argue Hartman and McKay, such sediments become buried in sedimentary basins, are never brought back into contact with the oceans and atmosphere, and are thus removed from active participation in oceanic and atmospheric chemistry. Because they are taken out of the system, oxygen can accumulate faster than when reducing compounds are constantly being reintroduced into the atmosphere, a case in which the dead don't stay buried. They make the intriguing point that Mars may have seen the evolution of complex life within a hundred million years of the formation of the planet (if one assumes, of course, that life originated there at all). Their argument is as follows: The rapid removal of reductants on Mars through burial in deep and undisturbed sediment would have allowed oxygenation to occur much more quickly than on Earth, where plate tectonics constantly recycles sediments through processes of subduction, plate collision, and mountain building. All these processes can cause previously buried sediments to be brought back up to the surface, where their reductants would once more bind whatever atmospheric oxygen was available. Hartman and McKay also point out that volcanicity on a planet, like Mars, that does not possess plate tectonics is much lower than on the earth; thus the amount of reducing compounds (such as hydrogen sulfide) entering the atmosphere-ocean systems on Mars from volcanic sources would have been much lower than on Earth.

Let us posit a Mars habitable for some hundreds of millions of years, with an oxygenated atmosphere near the end of it. This could be ample time to get beyond microbes if there were life on Mars. While we would have no hope that such life could still exist (in contrast with the case for microbial life, which might still persist on Mars), the fossil record of Mars is another story. Even if life made it only to the simplest form of multicellularity, we would probably still find evidence of this in the Martian sedimentary record. For this reason alone it is imperative to send a paleontologist to Mars someday.

What kind of alien life might do best on Mars?

Mars of today in no way resembles Mars of the deep past, just as Earth of today in no way resembles ancient Earth. Four billion years ago Mars may have looked like Earth of today without the mountain chains and other relicts of plate tectonics. Or maybe it too had plate tectonics and lost it, so perhaps it was more of a match for the earth of today than we think. But Earth back then was much more water-covered than now, perhaps completely so. But the key point is that Mars has undergone a far more fundamental change than Earth has. Life originating on Mars would have had earthlike conditions as a starting point. Mars of now is hostile to the early type of life and mostly hostile to present earth life. The best kind of life for Mars of today would be something that can stand cold, no oxygen, and little or no liquid. It is hard to imagine any kind of CHON life tolerating the place, or silane life, or any of the varieties of things that we have described in chapter 3.

Perhaps, just perhaps, the life that we find on Mars is ribosan, RNA life. That would be sweet. Most probably, if there is life, it is some kind of methanogen. But with what kind of genome? Is it Terroan life or Mars life, evolved on its own? Or are we Mars life?

Life of the inner planets

Does Mars have life? I repeated this question to Don Brownlee on September 25, 2004, after a new spate of results showing a match of methane and water vapor was announced. Don is above all logical. His answer is that the simplest explanation is that we have detected the presence of life on Mars. He thinks that we have found it. It is now time to send up not rovers but sensors specialized to find life and to bring it back to Earth.

With this chapter we leave the warmth of the inner solar system and pass outward to areas where even Mars seems like the tropics. In the next chapter we look at the first of the ice-covered bodies. Next stop: Europa.

Chapter 11

Europa

The innermost satellite, he [Galileo] insisted, should be named Io after one of Zeus's most beloved maidens; the second would be Europa, a Phoenician princess abducted by Zeus and later the mother of three of his children; the third would be Ganymede, a Trojan boy whom Zeus made cupbearer to the gods; the fourth and last would be Callisto, a nymph loved by Zeus.

—Jeffrey Kluger, *Moon Hunters*

The story behind our current fascination with Europa as an abode for life is a tale of submarines and spacecraft. In the last quarter of the twentieth century, two groups of planetary scientists sent ever more sophisticated robots and manned vessels in opposite directions. The 1977 discovery of hydrothermal vents in the Galápagos Islands, and the follow-on dives into an even more astonishing undersea volcanic system in 1979 off Baja California (dives profiled in chapter 1), gave an enormous boost to the possibility of life in space, even as NASA was sending a series of robots toward the moons of Jupiter. This convergence brought about a rapid realization and acceptance that Europa, a frozen moon of Jupiter's, harbored a deep and dark ice-covered ocean, with secrets as well kept as the dark and energetic sea bottoms of the earth's own deep oceans. Like all environments that might hold life beyond the earth, we must have two crucial questions:

Europa from space. Image from the NASA Galileo *spacecraft.*
(Courtesy NASA)

Are there reasonable means and conditions that could bring about the formation of life (if one assumes that it does not get there by panspermia)? And once life exists there, could it maintain a viable existence for long periods? The challenge for life on (or in) Europa comes down to whether or not life anywhere can evolve and then exist in an environment cut off from light. If not, then only worlds with sun-warmed surfaces can harbor life, and we will have removed a huge number of potential planets and moons as places to look for life in the universe. These, then, are the stakes involved in the search for life on Europa, for it contains an enormous but dark sea under a moon-girdling cap of ice. In this chapter we will examine Europa as a birthplace and then habitat for life—warts and all.

The unexpected moon

In any small telescope—indeed, even with a good pair of binoculars—the four dots swarming around the giant planet Jupiter are starkly visible. They are but pinpricks, but even over several hours they can be seen to be slaves of the giant planet's gravity. Humankind has known of them since the time of Galileo, but they have been nothing to us but points of light. It took the space age to put any sort of face on these distant celestial bodies.

Our first insight into the nature of these so-called Galilean satellites of Jupiter came in 1979, when two *Voyager* spacecraft passed through the Jovian system. After years of travel they spent a total of only two days there, but as they tore through at high speed, they took thousands of photographs of moons and planet alike, sending the information back to earth as faint coded radio signals. The unscrambled signals revealed unforeseen wonders: the cold and ice-covered surfaces of Callisto, Ganymede, and Europa, and the seemingly hellish and pimply Io. Each revealed an unexpected face. No one had predicted the active volcanism on Io that was observed or the strange cracked and grooved surface features of Callisto and Ganymede. But it was Europa that may have provided the greatest surprise. Where craters were expected, few were seen. It seemed to the scientists at the Jet Propulsion Laboratory in Pasadena, who were frantically downloading the images, that just like Venus, which had its craters and melted them away through planetary resurfacing, Europa was somehow erasing the evidence of its deep past. But where Venus accomplished the trick by melting everything on the surface to slag, Europa is very much a world of ice. It was not through flows of lava that Europa changed its face, although heat of a different kind is surely implicated. Europa looked more like ice-capped ocean in the polar regions of our Earth than a faraway moon halfway across the solar system, and the most conspicuous geomorphology was a complicated pattern of crisscrossing lines and ridges.

David Grinspoon, who was in the room when these pictures rolled in, quotes Carl Sagan, also there, as saying, "Percival Lowell was right. Only the canals are on Europa!"

The *Voyagers* passed through but left a lasting impression on their builders back on Earth. Europa was fleshed out. It is not large, being the smallest of the four main moons (at 1,945 miles wide). It is a rocky place (unlike giant Jupiter, a huge gas ball) with no atmosphere to speak of. But the most astounding aspect is its ocean. The surface is covered with ice, all right, but the sea it covers might be as deep as 150 miles, sitting atop a core of silicate rock. Unlike our core, which is dense and laced with radioactive elements that provide a constant flow of heat outward, Europa's core is no denser than granite, and just as cold. Hence the excitement and puzzlement over the surface. How could resurfacing of the ice cover take place? Was it periodically remelting, but if so, how? Without heat flow or volcanism, there seemed no reasonable explanation for the observed surface. Could large asteroids crashing into the planet occasionally melt the ice? And how thick was the ice? This latter question was of paramount importance to some, for once it was ascertained that the ocean of Europa was made up of water, the excitement began. Where there is water, there is the hope for life, earthlike life, in fact. (NASA has always had a single-minded plan for its pursuits of aliens: Follow the water.) So how was the surface of Europa getting its occasional face-lift? The answer, perhaps not surprisingly (because of its nearness to Jupiter), was that the gravitational effects of the giant planet that they orbit heat both Europa and Io. Their distance readily explains the relative degree of tectonic activity on each of the four Galilean satellites from Jupiter. Of the four, Io is closest to Jupiter and the most active tectonically. Like Europa, Io cannot hold the heat flow from radioactive minerals in the moon's core responsible for its volcanism, which is actually caused by the powerful gravity of Jupiter's squeezing the planet with such force that it has become a volcanic inferno. The crushing gravity working on Io as it revolves in close orbit generates enough heat to produce the magmatic eruptions, the same sort of heat generated when you rapidly bend a thin piece of metal back and forth with great speed and force. The consequent eruptions are frequent and violent. Huge gouts of lava of up to ten thousand tons at a burst are blown out of the many volcanoes. It is not only thermally vigorous but electrically charged as well. Giant Jupiter generates an enormous magnetic field, and as poor Io orbits through this field, it develops a huge charge of electricity—as much as one trillion watts constantly discharging through the thin atmosphere in the form of blazing sheets and pulses of lightning. There is no water on Io, just roiling rock and fulminating energy. This is not a place where we could imagine any life at all familiar to us.

Europa is next, but much farther from Jupiter than Io. Europa is almost twice as far from Jupiter as our moon is from us. Nevertheless the same gravitation flexing that tortures Io also affects Europa and does produce heat, but far less than on Io. In the parlance of the story of the three bears, the gravitationally generated heat might be "just right" for life, enough heat to produce, perhaps, underwater hydrothermal vent systems that could be the sites of life's origin and habitat on the deep, cold Europan sea bottom.

Farther out in space, Callisto and Ganymede get no heat from gravity and are frozen, dead balls. Thus, if Goldilocks exists in the Jovian system, she will be found, perhaps cold and sleeping, on Europa—in the form of microbes if there is life at all. Jack Cohen and Ian Stewart (authors of the recent book *What Does a Martian Look Like?*) plan to hunt for giant jellyfish and other animals in the Europan Ocean, and the great Freeman Dyson once actually suggested looking for freeze-dried fish, splashed off Europa from impacts, that might now be in orbit around Europa! (I have never been sure how much of his tongue was in his cheek when he let fly with that interesting vision, fish in orbit around Europa. With the intense radiation to be found in this neighborhood they might be already cooked.) Fun aside, let us hope that even something as complex as a bacterium or an archaean or something else that is native will be found there. But how much hope is there?

Comparing Europa with Earth

For comparing Earth and Europa, the most important variables appear to be temperature, radiation, pressure, energy sources (or, to a chemist, we can refer to these as redox gradients), and salinity. We want to look at three potential Europan habitats: the surface ice, the subsurface sea, and possible deep environments near hydrothermal vents. How will these environments stack up against earth conditions, and could we expect there to be life with earth tolerances in any or all of these environments?

There has been enormous debate about the thickness of the Europan ice cover and whether or not there is communication with the liquid water beneath it on some regular basis. The correct answer is crucial, because it has a great bearing on both pressures that would be encountered by any Europan organisms, as well as nutrient flux. Chris Chyba, one of the world's experts on Europa, puts the ice and water ocean at 80 to 170 kilometers thick, with only the top 10 kilometers being ice.

The amount of nutrients available is at the heart of a debate about the potential for life in the Europan Ocean. Life, as we have seen, utilizes oxidation-reduction reactions to extract energy. To maintain these reactions, there would have to be a cycling of oxidants and reductants. The surface ice of Europa is surely strongly oxidized because of the tremendous radiation levels there, whereas the open sea and sea bottom on Europa must be reducing. Only if there is a cycling of these two environments might we expect life to be common or at all present on Europa. For that to happen, there must be some way that the ice can be melted and its contents brought down into the liquid parts of the ocean. Thus the ice thickness is of paramount importance.

Observations by the *Galileo* Jupiter probe of the mid-1990s resulted in new information that gives more hope to those looking for life on Europa. One of the most intriguing observations was of large yellow-brown discolored areas on the surface of Europa. To the hopeful, this was an indication that the icy surface contained suffi-

Close-up of ice-covered surface of Europa, showing the numerous cracked areas where water might be reaching the surface. Photo from NASA Galileo *spacecraft.* (Courtesty NASA)

cient minerals and perhaps organics that life might form there or, indeed, that the discoloration was life itself. But the probe also allowed new estimates on the turnover time of the ice cover. Because it had few craters, it did not date back to the origin of the moon or anywhere close to that. Although the estimates were crude, Chyba recently concluded that the amount of time between complete resurfacing of the ice cover is on the order of ten million years. Such a long-term interaction between ice and underlying sea would spell doom for any sort of Europan ecosystem. But the surface of Europa shows a complicated terrain of ridges and grooves, leading planetary scientists to suspect that while much of the ice cap is indeed very thick, there are parts that might be much thinner, perhaps no more than a single kilometer thick, and that in these thin areas there may be areas where interaction between ice and sea is much more dynamic than beneath the thicker areas.

This gives hope for life. But even the most optimistic admit that the chance of ecosystems driven by photosynthesis is remote. The best hope would be for photosynthesizing microbes in the surface ice. It must be remembered that this far from the sun, photosynthesis would be a stretch even under ideal conditions, and even shallow burial might reduce sunlight to levels making it impossible. The trouble with living on the surface of the ice is the great flux of incoming and lethal radiation. While this appears to be a curse, there may be a blessing for life, but life that might be quite alien compared with ours, at least as far as metabolism is concerned.

By the late 1990s these discoveries had generated great enthusiasm for the possibility of life on Europa. In 1999, for instance, the National Research Council released a report suggesting that Europa was as important a target for the NASA search for life as was Mars. Ron Greeley, a council member, was quoted as stating: "Europa could be a better target for exobiology than Mars." This sanctioned enthusiasm, coupled with the discovery of the under-ice sea, led to lots of interesting speculations. For example, here is a passage from Cohen and Stewart, *What Does a Martian Look Like?:* "[T]he [Europan] ocean would be full of bacteria, with filter feeders and grazers [thus animal life] on the rocky seabed. There could be fishy forms, streamlined feeders eating the grazers and each other with shapes similar to Earthly fish, because shape is constrained by hydraulic swimming constraints." So much for the optimistic view. But near the end of the century a more sober assessment was made that, if correct, might spell the death knell for hope of life not only on Europa but also in any ice-covered ocean. Eric Gaidos, then a postdoc with my long-term colleague (and now University of Washington team member) Joe Kirschvink, published an article with Kirschvink and Ken Nealson that explored the possibility of energy sources under the ice. They were not looking at what kind of life might be there or how it evolved, but at whether there was sufficient energy to keep an ecosystem alive in the cold black. Their answer was a resounding no. The authors pointed out that practically none of the energy sources available to earth life would be available to any life under the Europan ice. Their analysis precludes any kind of multicellular life (Good-bye, fish! Good-bye, filter feeders!) and might even be insufficient to power microbial life. If any life does exist there, it would be at such low concentration as to be undetectable by any imaginable sensor—short of a human with a microscope and a lot of patience.

The argument mounted by these astrobiologists is critical. In essence, they are saying that CHON life needs more than water, hydrocarbons, and other chemicals. It needs a source of energy, be it solar or geothermal, that maintains disequilibrium and powers oxidation-reduction reactions. Compared with Earth, generous estimates of the European chemical energy available to drive a biosphere are only 1 part in 100,000,000,000,000 per m^3 of water. Ouch! Little energy (and this estimate is of a very small amount of energy compared with Earth), little life.

So there are huge caveats. Let us look in detail at the three potential habitats for life: the ice cover, the open sea beneath the ice, and the sea bottom.

The surface ice layer

One of the most powerful and influential movies of all time was Stanley Kubrick and Arthur C. Clarke's *2001: A Space Odyssey*. Much of the movie takes place in the environs of Jupiter, and we see two humans (and one crazed computer) spending some quality time with one another in the Jovian system. While the movie still has not been surpassed in its rigorously correct depiction of space (ever since, we have been assaulted with noisy spaceships in the vacuum of space, among other sins), there is one great flaw in *2001* that was unknown at the time to its creators. Jupiter is a neighborhood that would be quickly lethal to earth life, human or otherwise, unless shielded by impractical (for spaceships) amounts of lead. Jupiter emits lethal amounts of ionizing radiation. Enough of this is present to make the surface of Europa a very dangerous place to be. Because of this, humans may *never* be able to travel to the Jovian system and

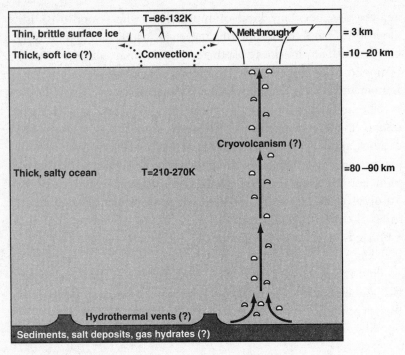

A schematic diagram of the surface layers of Europa.

return to tell about it. How lethal is it? A common bacterium would be killed with a ten-minute exposure, and even a bacterial spore, highly resistant to all kinds of environmental stress, would be inactivated in forty hours. Anything *on* the ice would be quickly killed. But *in* the ice is another matter.

This radiation is caused by the huge magnetic field generated by Jupiter. Particles are trapped and accelerated by this magnetic field, and unfortunately for Europa, the huge field and its whipped-up radiation extend far beyond its orbit.

So the radiation is lethal for earth life, but could there be life that takes advantage of the radiation as an energy source and avoids being irradiated to death? And could there be any sort of earth organism that might use the radiation to its advantage? According to Chris Chyba, there is. Chyba was Carl Sagan's last disciple, and early in Chyba's career, the two modeled potential energy sources for life on Europa. In a recent article Chyba and his coauthor, Cynthia Phillips, further elaborated how this might work.

The great flux of energy hitting the ice on Europa will have very distinct chemical effects. While radiation hitting salt ice will do little, the salty ice covering Europa will also have the dispersed products of the numerous comets that have crashed into the ice and thereby spread their loads of organic compounds and metal far and wide. The radiation hitting the ice will cause these products to oxidize and thus could provide a food source. Of course, if they continue to stay on the surface, they will be destroyed, but even a slight burial will shield them from further radiation "poisoning," as it were. (This same radiation is actually sputtering the surface of the ice off into space; hence very little of the oxidized materials will be left for mixing with the ice.) As on Mars, an expected by-product of the radiation will be hydrogen peroxide. This can be a source of free oxygen, which can interact with the organic molecule formaldehyde (CH_2O) to power organisms by the reaction of formaldehyde and oxygen reacting to form water and carbon dioxide. Just this reaction is used by a soil bacterium on Earth. If these organisms existed in the top layers of the ice on Europa, they might thrive in virtually untold numbers. So here is an earth life-form that, if let loose on Europa, could survive the radiation and even get a meal, according to our best guess. However, there is another factor, temperature. It turns out that the surface ice layer is anything but tropical.

It is really cold at the surface—the temperature there varies between $-304°F$ and $-228°F$, which is way, way colder than the minimum temperature for earth life (about $-20°F$). At these temperatures, ice is as hard as steel. Only at the base of the ice layer would we find temperatures that could support life as we know it, and the Chyba hypothesis requires that life live in the upper layers only. If that were not enough, the intense cold of the surface would be so strongly desiccating that life would quickly perish. So this is an impasse for life wanting to live on the surface; it would need a radiation suit, a heater, and a humidifier. Unless there is some form of really cold life, the ice will be sterile down to near its contact with the underlying seawater.

Just being able to live is not enough. Could life have evolved on Europa? Perhaps life cannot find a way on the surface of the ice, but there may be at least hope for the synthesis of organic molecules in surface cracks. The amount of sunlight getting to the surface of the ice is only 5 percent of that hitting our earth surface, but according to work by Jonathan Lunine (another University of Washington NAI team member) and Ralph Lorenz, this might be enough energy to make organic molecules. This could be of great use in the formation of potential life on Europa. But could life form in such cold? Certainly not ribose-based life, and therefore we Terroans, as the calcium-borate minerals necessary for the formation of Terroan life (at least using the recipe of Steve Benner that we read about in chapter 5), would dissolve in the oceans. So, very little hope for life as we know it at least evolving here, using the recipes that we think we know.

We will beg that question for the moment and continue our descent into the cold black waters of the Europan Ocean. Could there be life there?

The Europan Ocean

One of the first oceanographers to grasp that Europa might be habitable was John Delaney at the University of Washington. Along with his UW colleagues John Baross and Jody Deming, Delaney was instrumental in creating the astrobiology group that ultimately morphed into one of the world's best centers of astrobiological study. Delaney had spent his career studying the deep hydrothermal ridges and mid-ocean ridge spreading centers off the Washington State coast. This was brutally hard work, done under taxing conditions in the wild North Pacific Ocean. (A good review of Delaney and his work can be found in William Broad's magnificent book *The Universe Below*.)

As Delaney pondered the nature of the Europan Ocean in the late 1990s, he concluded that it would probably be stratified, much the way our own planet's oceans were during long periods of the Mesozoic era. Today our oceans are mixed, meaning that they are fairly homogeneous in chemistry (including oxygenation) from top to bottom. But a stratified ocean might be an advantage to life, at least on Europa, for as it stagnates, it sets up temperature gradients that can lead to overturning and mixing upward. This might be the mechanism that brings reduced matter back into surface regions if that happens at all. Delaney became a major proponent of vigorously pursuing the search for life on Europa.

Any life will be at the mercy of the temperature, salinity, acidity, and pressure of the Europan Ocean. Like our Earth's seas, the Europan Ocean is saline; this is not a great freshwater lake. There is some controversy on just how salty it is and what the chemical constituent of the brine might be. Because the composition and concentration of the brine minerals will have an effect on the acidity of the sea, the exact nature of the brine is of prime importance in deciding how welcome Europa's ocean is to life. Three possibilities exist: It is a neutral mixture of sodium and magnesium sulfates, it is an alkaline solution rich in sodium, or it is an acid solution with sodium and magnesium. Each possibility has an effect on the kind of life that might be present (or on whether it is present at all), but barring some instrument actually getting a sample, we may never know. Life in the form of extreme halophiles (salt lovers) could conceivably survive the first two models, but the latter, the acid sea, would probably be inimical to any sort of life.

All in all, the Europan Ocean would present some opportunity and plenty of challenges for life. What about the bottom of this sea?

The sea bottom

The last potential site for life on Europa would be its sea bottom. Because of its extreme depth, this would be an environment far more pressurized than the deepest ocean on Earth. If the Europan Ocean is 100 km deep (about the middle value of the estimates for its depth), the pressure on its bottom would be 1,200 atmospheres, re-sulting in a pressure of 17,640 pounds squeezing every square inch of any poor bug living in the deep, dark, pressurized cold of the Europan sea bottom. This is way in excess of the pressure found on the deepest terrestrial ocean bottom. The Europan sea bottom seems to be a very poor place for life *unless*—and this is a huge unless—it turns out that there are hydrothermal vent systems down there. This might be the best hope for life in the Europan Ocean, since hydrothermal vents can cause great rises in ambient temperature around them. They are also places where organics and energy supplies are to be found. The evidence that such systems exist comes largely from models of the ice disturbances at the surface; these might be caused by upwelling waters of warm temperature created in deep hydrothermal vents.

New calculations by a University of Washington astrobiology grad-uate student may have further dampened hope for life on Europa. Steven Vance looks at the way the fluid packages move about in the Europan Ocean through mathematical modeling. But another part of this work is looking at the chemistry of the ocean. Vance postu-lates that the greatest depths of the Europan sea are saturated with magnesium-rich salts. The concentration of this salt is so great that there may be no way that any earth organism could live there. And could any alien biochemistry deal with large salt concentrations at the high pressure of the European sea bottom? This is one more rea-son that I am not optimistic about life on Europa anywhere.

But what about energy?

Earlier in this chapter I discussed the 1999 work of Eric Gaidos and his colleagues that suggests there would not be enough energy to power aliens on Europa. If that study stands, it is unlikely that we will find life anywhere in the Europan system; even if it evolved, it would not have sufficient energy to live. But that view has not gone unchallenged, and in 2004 another take on the subject was published, by the same two astrobiologists who have speculated that there might be life in the clouds of Venus: Louis Irwin and Dirk Schulze-Makuch. Why have they taken this opposite point of view?

This new view on the amount of energy available to any possible Europans really derives from a single estimate, made by Tom Mc-Collum, in 1999, on just how much energy might be liberated by taking advantage of the hydrogen and carbon dioxide being vented from the deep mantle of Europa. The gravitational tugging of giant Jupiter on Europa, as we have seen, is a source of warmth and energy. Deep inside Europa the gravitational effects on the stony interior probably cause a volcanolike result, the release of carbon dioxide and hydrogen, which are also produced by terrestrial volcanic events and our Earth's mantle as well. Free hydrogen is a wonderful energy source, and carbon dioxide yields all the carbon any organisms could want. This very system is used by a huge and important group of archaeans on earth, the methanogens. By mathematically modeling the release of hydrogen and CO_2 from the Europan mantle and by assuming that some fraction of this is available to microbes—if that they are there at all—the authors arrived at a rough estimate of the biomass that could conceivably be found in the Europan Ocean at any given time. Assuming that "only" 10 percent of the available energy was utilized by the chemolithotropic Europan microbes, Irwin and Schulze-Makuch estimate that at any given moment—this moment, for instance—there are about a hundred tons of primary producers (the microbes that take the hydrogen and turn it into living cells). If we assume that evolution has produced species that can use dead cells and other organic detritus for energy (cells that eat both of these), there might be an additional ten tons of each. One hundred twenty tons of living cells in an ocean with twice the volume of that found on Earth is pretty paltry. We could go there and never find them. The authors also computed potential biomass if the microbes used not hydrogen but ionizing radiation as an energy source. Under this scenario a slightly higher biomass was computed. If this second group of organisms existed, they would be aliens; no such life exists on Earth.

What kind of alien life might best survive on Europa?

So what are the hopes for terrestrial life surviving in the oceans of Europa? If the pH is >11 (thus very alkaline, like ammonia), as many planetary scientists now suspect, and if energy supplies are low, the odds are small; the energy needed to maintain an "earth-normal" cell in such high ammonia and alkaline conditions are probably greater than the energy available.

The best hope for life on Europa might indeed by something very alien to earth standards. Unless there are hydrothermal vents on the sea bottom, there is very little hope that any earth life could exist there. While those interested in life on Europa hold great hope for life in the surface ice following models of Chyba and others, the review of Giles Marion and his colleagues points out the extreme difference in temperatures between the surface ice temperature on Europa and what any earth life analogue could survive. We are not even close here. Ice is a very poor insulator; any microbe wrapping itself in ice or even mucus, as extreme cold earth life does, will soon freeze to death.

What biochemistry might work? There are indeed alternatives. Biochemists have suggested a form of life informally dubbed "ammono life," and in this there would be a chemistry quite different from the earth life chemistry. For example, phosphate, an essential part of the energy-gathering and metabolic system of earth life, would have to be replaced, since complex phosphate-bearing molecules, such as ADP and ATP, would be quickly destroyed. Our ammono life would have to use molecules that are stable in ammonia and water, can hold multiple negative charges—a requirement mentioned by Steve Benner and his colleagues—and can form stable bonds that are amidelike with carbon molecules. With such a system a wholly different form of energy transfer would have to take place. Dissolved metals and their liberated electrons could become the energy gradient, producing a form of energy transfer similar to that seen in electricity, where electrons pass from place to place. This is very different from the terrestrial system where positive-charged protons are the currency transferred from ion to ion, such as the addition and subtraction in the ADP and ATP pathways that we use and that yield energy. This is probably the most important point here and the real reason to look for life on Europa, according to my colleague Joe Kirschvink.

How and where to look for life on Europa

Since for both political and biological reasons it is unlikely that humans will be landing on Europa anytime soon, what sort of instrumentation should be included so as to avoid the ambiguity that was the *Viking* life detection equipment—not being able to detect life even if it is there? Clearly, the probe that lands on the ice surface need not be equipped with a hole cutter and an ice fishing rod and reel. Since there is no source of free oxygen under the ice, there will be no Europan trout or pike to haul out; the chance of animal life is minimal. So how to test for microbes, and where? Chyba and Phillips suggest a landing near the most jumbled terrain (this will surely make the NASA managers in charge of a safe landing turn prematurely gray), because such a site will have the best chance of rapid exchange between the ice and the ocean beneath and thus the best chance for life. It is also the worst place for a spacecraft to land safely. There have been areas suggested to be transient Europan "ponds" that might be optimal for the search for life. There should be a coring device so that the seawater underneath can be examined. Some optimists hope for a submersible of some sort to explore the subsea, but such a mechanism would have to support high pressure and ultimately have to surface again so as to transmit its information. That is asking a lot of a robot. One thing is sure, however: If such a mission is undertaken, we had better be sure that we do not infect Europa with our own life. The only panspermia in the solar system might be that created by the various space programs, but it might be significant. Have we already infected Mars and Titan with earth life? Some people in NASA are afraid that we have. I will return in detail to this problem near the end of this book.

Europa of the past

If Europa seems a pretty cold place now and thus not at all a hopeful place for earthlike life, what about in its deep past? New work by NASA scientists suggest that like so many other places in our solar system, Europa long ago may have had a liquid ocean at its surface, as opposed to its current icy crust. If that was the case, the possibility of life's forming is less remote.

Imagining a Europa warmer than now seems counterintuitive. Long ago, in the first interval of our solar system's existence, the sun was dimmer than now, and even the paltry warming that far-off Europa receives from the sun would have been far less. But back then a body far more important to Europa than the sun, nearby Jupiter, would have been far more energetic, according to NASA Ames scientist Kevin Zahnle. The source of this warming would have been infrared radiation emitted by nearby Jupiter. It is possible, according to Zahnle's calculations, that Europa was warmed sufficiently to have an entirely liquid ocean, stretching from pole to pole, a real Waterworld. If so, there would have been an atmosphere, and with an atmosphere, life could have reached here, intact, by the panspermia systems of meteorite travel.

The odds

If I were a gambling man, I would not bet the house on finding life on Europa. It is too cold for good CHON life, but not weird enough, seemingly, and without enough energy sources to build a really good, really weird alien. Plenty of experts out there will howl that this is not true at all, for the true believers really believe that Europa is the place to go for a great on-location alien movie using real aliens, even if they are microbes. My guess is that we have to move much farther into space to find alien life, so on to Titan, the giant moon of Saturn and the subject of the next chapter.

Chapter 12

Titan

Only the prospect of life—indigenous or otherwise—would really justify a human presence on Titan.

—Ralph Lorenz, *Lifting Titan's Veil*

It is perhaps the most enigmatic moon in the solar system. If it were directly orbiting the sun, instead of Saturn, Titan would be considered a planet, for it is as large as Mercury. Yet it is a moon, not a planet, and one cloaked in a thick atmosphere that obscures its surface features—the only moon in our solar system with an atmosphere, in fact. What else does Titan hide? Many in astrobiology think that it may hold clues to the nature of the prebiotic earth. Others are bolder and speculate that after the earth and perhaps Mars, it is distant Titan that has the best chance of having life. In this chapter we will look at this distant Saturnian moon and concentrate on new findings from the *Cassini* spacecraft, launched in 1997 and arrived at Saturn in 2004. Is there life on Titan?

Much became clearer about Titan in late October 2004, when *Cassini* passed within several hundred miles of the surface. Photographs from this close encounter showed distinct black areas and whiter areas. But the main surprise was the near absence of distinct impact craters. The surface of Titan must be active.

In December 2004, the second part of the *Cassini* mission, the

Huygens probe, parachuted onto the surface of Titan and for a few precious hours transmitted much tantalizing information about the surface of this moon. Let us portray what the surface of Titan, the home, forever now, of *Huygens*, might be like. Where does the probe now rest? From the data that have been received, we can picture the gloomy surface of Titan. Let's imagine we are there.

We are standing on a large plain, ankle deep in dark, oily muck. Overhead, a huge and ghostly disk, the specter of Saturn, cutting through a dim orange haze, fills a large segment of the sky. However, there are no rings to be seen, for from Titan the iconic Saturnian hallmarks are edge on and far too thin to be visible; still, their shadow leaves a darker mark on the roiling Saturnian atmosphere.

Radar image of Titan taken by NASA. Life could be on the surface of the ice-covered ocean, in the ocean, or on the bottom. (Courtesy NASA)

We descend into the palisade-like ring wall of a large impact crater on Titan's surface. With the tiny sun now below the horizon, the thick, ruddy smog that is Titan's atmosphere darkens into dim magenta and then loses all color. Yet there may still be some light. Blue-white flashes from sparking lightning in the lower clouds and a yellowish light reflected from giant Saturn itself give a faint twilight illumination to the Titanian surface. In this dim gloaming a thin organic rain of dark methane is falling. In the crater we have entered a phantasmagoric wonderland of hydrocarbon fluid, ice, sludge, and rock, all arrayed in frozen glory. But the landscape shows motion as well, for thin streams of liquid ethane wander through the tar pit landscape. At a temperature of −180°C the steaming geysers of methane venting from the warmer interior instantly freeze and then fall as black organic snow while other organic liquids are belched out from a nearby fumarole. A larger river of methane snakes over the crater floor in the distance and finally disappears behind the back of the central uplift peak at the center of the enormous crater. The question is: Amid this frozen landscape is there life? Many of us in astrobiology think so. At worst, Titan can teach us about how life first evolved on Earth. At best, it is a veritable zoo of enormously alien life-forms. We need to go there. We humans as well as robots. Could there be greater adventure awaiting our species? I doubt it.

The *Cassini* mission

On October 15, 1997, a Titan/Centaur rocket blasted off from Cape Canaveral, Florida, to begin a seven-year voyage out from Earth. Its destination: Saturn. The payload was the six-ton *Cassini/Huygens* probe, and a great deal of nervousness by both NASA and a vocal segment of the interested American public who paid for the mission accompanied the *Huygens* spacecraft and its takeoff. The successful insertion of the large spacecraft into orbit was met by a considerable sigh of relief from all concerned, for *Cassini/Huygens* carried a plutonium-containing reactor, and a repeat of a shuttle launch disaster could have sprinkled lethal quantities of plutonium over parts of Florida and its coastal waters. Happily, off it went into space without a hitch, accompanied by a relieved "I told you so" from NASA brass.

The name of the two-part spacecraft—*Cassini/Huygens*—is apt. Giovanni Cassini discovered the larger moons of Saturn now known as Iapetus, Rhea, Dione, and Tethys, as well as the breaks in Saturn's rings, now known as the Cassini divisions. Christian Huygens discovered the largest of the moons, Titan, in 1655. Appropriate, then, that the larger part, the *Cassini,* was to survey these larger moons and Saturn itself, including its rings, while the smaller *Huygens* part was designed to parachute onto the surface of Titan itself, as it did January 14, 2005.

Now there is a new crew of scientists interested in this enigmatic moon, and one man in particular has guided the scientific exploration of Titan. He is Jonathan Lunine and he is a member of my University of Washington NASA astrobiology team, I am happy to say.

In a recent interview, Lunine summarized the three great questions about Titan: First, why does it have such a thick atmosphere, when the similarly sized Jovian satellites Ganymede and Callisto have no atmosphere at all? Second, where does its methane come from? Finally, how far toward life has it progressed?

The *Huygens* probe was released from the *Cassini* orbiter, and on January 14, 2005, it parachuted onto the surface of Titan. Its primary mission was to study the atmosphere of Titan as it slowly drifted down through the thick atmosphere. Its pictures of the landscape and surface exceeded all expectations.

The curious atmosphere of Titan

Titan is beloved of NASA for many reasons, but one of them is purely practical: It is one of the only targets that NASA covets where a spacecraft can land by parachute. Titan's atmosphere is thick and, for those searching for life in the solar system, very, very tantalizing.

It has a pressure at the surface about that of the earth's. But it is not so much the thickness of the atmosphere that intrigues as the composition of Titan's air, which has kept planetary scientists scratching their heads for decades. Early on the chief scratcher was Carl Sagan, and it is safe to say that he went to his grave mystified by the ruddy gas above Titan. In the 1970s there was an understanding that Titan's atmosphere was thick, but there was no way of really identifying all its constituents beyond the discovery by earth-based spectrographic telescopes that there was both hydrogen and methane in the atmosphere. The discovery of methane was the most exciting (just as it has been for Mars, and for the same reason: Is it from life?), and with this discovery Titan moved up on everyone's list of places that could conceivably harbor or that once had had life. A far-distant and previously unpromising sphere of very cold real estate had shown an organic molecule in its atmosphere. But the analyses were necessarily crude, and little beyond the identification of these gases could be made.

Even at this level excitement was mounting. Methane does not last long in any atmosphere; when hit by sunlight, it breaks down into hydrogen atoms and lighter molecules that then readily combine with others to form longer and more complex organics, such as acetylene and ethane. The most curious aspect of all this was the presence of the methane itself. Because it does not last overly long before photolysizing, there had to be a source of methane generation on the surface of the moon itself. While there are several inorganic ways to do this, the easiest way it is done, and how we get methane in earth atmosphere, is by the action of microbes. As we saw in chapter 10, methane can be a signal of life's presence. But the alternative—that it comes from volcanic activity of some sort—is also exciting.

While much could be measured and inferred from spectroscopic looks at this far-distant moon, it was not until the *Voyager* arrived that an accurate assessment of the thick smog surrounding Titan could be made. These results showed that most of the atmosphere there is nitrogen, followed by methane, and then argon. But once again this discovery led to further excitement about the prospects of life on Titan.

Nitrogen combines with organic constituents to produce nitrogen-containing organic molecules. This is a stairway to life.

By the 1990s there was a pretty good idea about the nature of the Titanian atmosphere. There are certainly many of the building blocks for life. But to build life, at least our kind of life, two other molecules are necessary: water and oxygen. Because of the frozen surface of Titan, both are in short supply now. But what of the deep past?

Titan past and present

An enigma of planetary science is that so many of the bodies in our solar system may have been warmer in the past—in most cases, much warmer. The paradox here is that as we go back in time, the sun was cooler. How could there be a cooler sun and warmer planets and moons? The answer is in the greenhouse effects of certain atmospheric gases, such as water vapor, carbon dioxide, and especially methane. These gases trap warmth against a planet, and even early in the history of our solar system, when our sun had only two-thirds its current energy output, there may have been many more planets and moons that could have been warm enough for liquid water. Titan has been deemed one of these.

Yet if we were able to go back in time, as well as journey through space and visit the Titan of three billion years ago, any life that had formed there (or had traveled there and taken hold through panspermian processes) could not have used sunlight for energy. Surely if life lived there, it would have been life existing on chemicals in the rocky surface in any oceans then present.

Ammonia ocean of Titan?

There are many ideas about what the surface of Titan may be like. But one of the most compelling hypotheses concerns not its surface, but its subsurface. An image of this might be seen in the present-day

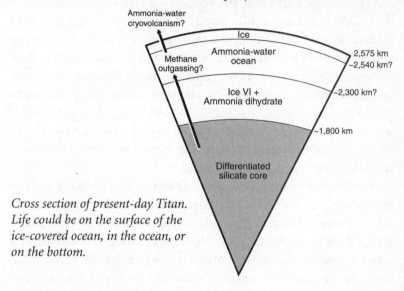

1.5 bar N₂/Ch₄ atmosphere

Ammonia-water
cryovolcanism?

Ice

Methane
outgassing?

Ammonia-water
ocean

2,575 km
~2,540 km?

Ice VI +
Ammonia dihydrate

~2,300 km?

~1,800 km

Differentiated
silicate core

*Cross section of present-day Titan.
Life could be on the surface of the
ice-covered ocean, in the ocean, or
on the bottom.*

North Pole just before the start of spring. A weak sun lifts above the
horizon but gives no warmth. The surface is thick ice. Yet beneath this
ice is ocean. On Titan there may be a similar cover—not of water ice
but of ethane and methane ice, while the ocean underneath is not water but ammonia with some water mixed in. Life, if it exists there,
would have arisen in chemistry radically different from that of Earth.

The formation of Saturn and its many satellites probably occurred
in a portion of the solar nebula (a great disk of matter that eventually
formed all the bodies of the solar system through particle-to-particle
accretion) that was ammonium-rich. Ammonia is pretty nasty stuff
for earth life; it kills microbes (hence its use as an antimicrobial agent
for cleaning kitchens and bathrooms), and its fumes alone can cause
tissue damage if inhaled in sufficient quantity by higher life-forms on
Earth. As it coalesced out of this mix of noxious chemicals, Titan, Saturn's largest satellite, would have differentiated into a layered moon of

concentric shells of matter, each with a different composition and density. As on planet Earth, the heaviest material would have sunk to the middle of the moon (the core), but Titan's core would have been far less dense than Earth's. Whereas our core is made of iron and nickel, that of Titan would be silicate rock and ice, and as such its density would be even less than that of our crust. There are major consequences of such a light core for a planet or moon: There will not be radioactive, heavy elements creating a steady flow of heat from the interior of the planet, and without a spinning iron core there will be no protective magnetic field surrounding the moon, integral in deflecting dangerous cosmic rays from space.

Understanding the formation of Titan was the passion of a young grad student in the 1980s. Jonathan Lunine and his adviser, David Stevenson of Cal Tech, calculated that the dense core of Titan would be overlain by a shell of pure silicate rock, with the outer layer a water-ammonia ocean, with about 15 percent of the ocean composed of ammonia by weight. This would have been the nature of the brand-new moon. But our gleaming new Titan did not last long in pristine shape. It was dinged almost immediately by a rain of cosmic junk—comets and asteroids adding rocky material to the low-density surface ocean—and in the process the inflow of a carbon-rich and stony class of meteors would have brought about chemical reactions producing a great amount of organic molecules, such as amino acids and more complex long chain or cyclic organic compounds. Early in its history the water-ammonia ocean, along with the underlying rock layer of its sea bottom, would have caused the liquid in the Titan ocean to have been reprocessed into something much more complex than H_2O/NH_4.

Would this early ocean have been liquid or frozen? There is a possibility that it was liquid, for at that time the atmosphere of Titan very likely was even thicker than it is today, and greenhouse effects may have kept the surface regions warm, wonderfully warm, in fact, for it has been estimated that surface temperatures in Titan's warmest phase may have reached 80°F. It very well could have been liquid—and perhaps warm liquid with abundant energy sources.

This state of affairs may have lasted for a billion years, and during this time all the ingredients for life—earth life—would have been at hand—everything but warm sunlight, that is—and with constant lightning and flexing of the interior of the planet, as well as the still-active flow of heat from the interior left over from accretion, microbes, had they arisen, would not have gone hungry.

Would any life have been earth life? In a place where ammonia is so common, could it have been an integral part of life there? This was proposed in 1995 by François Raulin and his group, with the novel idea (that we explored in chapter 10) of what they called ammono life, where the organic molecules used by the formation of first earth life are different on Titan, utilizing organic where oxygen atoms are replaced by nitrogen/hydrogen chemical groups. *This* would be alien life. If we assume that it could have arisen at all, could it still be alive today? And what would it be like?

Titan life

No life has obviously been found on Titan. No human-made object has gotten there. But if we assume that life is there, or was there, what can we say about it with any confidence? In a thoughtful review in 1999, Andrew Fortes of England summarized our best guess, and the following is adapted from his article. Fortes showed that quite a bit could actually be said about possible Titan life. First, we know that there is no free oxygen, so any life would be an obligate anaerobe; it would live without using oxygen for life processes and in fact would probably be poisoned to death if it encountered oxygen. There is also very little sunlight, and if we go back in time to the early solar system, even less than now, 30 percent less. Does this preclude the possibility of photosynthetic organisms? Perhaps not. Calculations on the amount of light theoretically penetrating the Titan atmosphere show there to be as much as that used by some photosynthetic life on Earth. Yet even if photosynthetic life did exist in the past, the present-day ice cover would have probably doomed it to

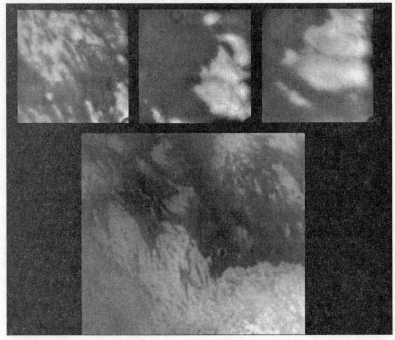

Cassini *images of Titan from October 26, 2004. The lack of impact craters visible in these high-resolution images suggests that the surface is slushy. The black may be ethane lakes; the white, ice. These images boost our interest in the potential biology of Titan.* (Courtesy NASA)

extinction, for after perhaps hundreds of millions, or even a billion years, the oceans of Titan would have frozen over.

Observation of Titan from earth-based and space-based telescopes is frustrating. Little in the way of surface features can be observed. But calculations based on an understanding of the chemistry present on Titan tells us that the surface is probably composed of water ice mixed with organic compounds or even lakes and rivers of ethane or methane atop the ice-capped ocean. But all the ice remains frozen, and liquid water would be available on the rare occasions when an asteroid or a meteor slams into the surface and melts the top regions of its crater. Recently there has been a good deal of interest in these transient "lakes" of freshwater. Suppose that an incoming comet or asteroid crashed into the surface, and the enormous heat melted the ice. For how long would the pool of water remain in a liquid state? Lunine and his students suspect that the water lakes might remain liquid for thousands of years. Would this be long enough for life to evolve? No one really knows how long it took earth life to evolve on Earth, but Titan is not Earth. We cannot discount the possibility that carbon life much like ours, or perhaps a variety better suited for Titan and its cold and very different chemistry, could get its act together and live—if only for a while—in these lakes. Eventually the lakes would freeze, but by that time would some microbes have evolved far enough to go into some hibernation stage that could come back to life with the next crashing comet?

Another aspect of this cold place is the very presence of oil and water. An old saw is that oil and water do not mix, as anyone unlucky enough to get water in his or her gas tank can attest to. This actually might work in life's favor on Titan. If an incoming asteroid melts some quantity of ice into water and splashes it as droplets into a pool of ethane or methane—some liquid hydrocarbon—the water forms small protocell-like structures in response. The converse is also true: Small drops of hydrocarbons falling into liquid water will form small protocells filled with hydrocarbons. For this latter case, is there any hope that life might arise? Possibly, according to those who have thought this prospect through. For example, Steve Benner and his group have postulated that the organic liquid could be the solvent for some sort of life. Organic reactions, the chemical reactions of life, work as well in organic solvents as in water, according to Benner. In fact, there are reasons to suggest that it might be easier to form life in this type of environment since, as we have seen, water creates its own problems for organic molecules. One of the bonds that can be used broken as an energy source is called a hydrogen bond. Forming one can store energy, and breaking it makes energy available. Creating compounds with hydrogen bonds is more difficult in water than in an organic solvent. Hydrocarbon life, if it used nucleic acids the way earth life does, would actually have some distinct advantages. We have seen that the synthesis of RNA and DNA in water is complicated by the tendency for water to deaminate the molecule, to destroy it by breaking apart the bases that code for information. This is always a problem for water-based life like ours, and Benner suggests that nucleic acids might like floating about in a hydrocarbon solvent much more than they do the water solvent of dear old Earth. Some hydrocarbons, like water, are polar, while others are not, creating situations where a hydrocarbon liquid becomes isolated from another liquid of some other chemistry even though the two are in contact. As we have seen, isolation is a necessity of life, and here is a way to achieve this.

For these reasons, the surface of Titan is of enormous interest to astrobiologists. It has the potential not only for life but also for several distinctly different varieties of life. Titan meets all the criteria that we think are necessary for life: There are plenty of carbon atoms already packed up in carbon-rich compounds, it has temperatures at which chemical reactions will take place (if slowly, in many cases, but who is taking away points for slowness?), and it has an abundant number of energy sources. The surface of Titan is definitely not in chemical equilibrium.

How do astrobiologists view all this? The following quote from Steve Benner gives a flavor of the most recent buzz about Titan: "This [the factors cited here] makes inescapable the conclusion that if life is an intrinsic property of chemical reactivity, *life should exist on Titan* [my italics; I find this sentence electrifying in its implications]. Indeed, for life not to exist on Titan, we would have to argue that life is not an intrinsic property of the reactivity of carbon containing molecules under conditions where they are stable." Wow. When we do go, in our spaceships with our collecting buckets and all the gear necessary to bring home life, we'll be possibly sampling the most diverse life—in terms of basic chemical types—to be found in the solar system, including Earth, perhaps life more alien than anything imagined by fevered science fiction writers.

Yet as bullish as the prospects for life on the surface are in some scientists' minds, there is even more hope for it below the surface of Titan. It is there where a monumental ocean of ammonia might exist, again with perhaps its own biology of alien life. So what might we find beneath the ice cap?

Life under the ice?

Oceans on Earth are relatively shallow. Yet as we have seen in chapter 11, the oceans of the outer solar system are a much different affair. Early on, the ocean underneath the icy surface of Titan (this ice cap itself is estimated to be about 15 to 30 kilometers thick) may be 700 kilometers thick, a nearly 500-mile-deep ocean by some calculations. Later it shallowed to perhaps 50 kilometers thick, according to Lunine and Stevenson, or as deep as 200 kilometers on the basis of the calculations of Olivier Grasset and Christophe Sotin. With a strange sea bottom of solid ammonia compounds and a smattering of rocky material coming from the fall of meteors into the sea before it was roofed over by ice, this would be a place even more alien than the ocean and its bottom on Europa. As the ocean shallowed, its ammonia concentration increased from the initial starting point of 15 percent to more than 30 percent by weight.

What would life have to deal with in this huge dark and (to us) poisonous sea? First of all, it is cold down there. Because liquid behaves differently under pressure from the way it does at earth sea-level pressures, it is possible to cool water to well below its normal (to us) freezing points. Also, by adding ammonia to the water, as on Titan, a really good antifreeze is present. While the lowest air temperatures on Earth can dip to −126°F, our oceans remain a good deal warmer, only just dipping below freezing at their coldest. (However, there are deep Antarctic lakes in which water temperature can reach −60°F.) But things are different on Titan. There the temperatures of the sea may reach −36°F. This is cold, but not so cold that earth life might be precluded. As Jody Deming and her students at the University of Washington have shown, such temperatures are not a barrier to life on Earth. And there may be hot spots in the deep Titan oceans where temperatures could reach a tropical 80°F.

If not theoretically barred by temperature, how about pressure? An ocean 120 miles deep, beneath another 15 miles of ice on top of the ocean, adds up to a great deal of pressure if one is a bottom dweller—about four times as much pressure as that encountered on our planet's deepest ocean bottom. Could earth life grow under such pressures? No microbes have ever been observed to grow at such pressures, but then again they would never encounter such pressure on Earth. Proteins would denature (be destroyed) at pressures of those in the deepest part of the Titan ocean, but only just, and it might be possible for earth life to live under such pressure. DNA can certainly take it, for it remains stable to pressures twice that found in the deepest ocean of Titan.

A greater problem than temperature and pressure might be the alkalinity of the ammonia-rich ocean. Ammonia makes solutions highly basic (the opposite of acidic), and the pH of the Titan ocean has been calculated to be −11.4. But here pressure helps the situation, for under pressure the pH is reduced to −10.5. Earth life could survive this easily, it turns out. Microbes have been observed to grow at a pH of −12, so Titan's ocean does not pose a problem on this score.

The next challenge is viscosity—the relative thickness of a liquid. Titan's oceans would be far more viscous than Earth's oceans. Any organisms would be swimming or floating in syrup. Again, this does not inhibit life, but it would reduce mobility and perhaps food acquisition. For a microbe, there would be little effect on life. There certainly would not be any world-record swimmers, but it is highly doubtful that anything has ever taken a stroke in the Titan ocean anyway. If life is there, it is probably a floater or a bottom slick.

How alien might Titan life be? Perhaps very. It has been speculated that in addition to an ammonia ocean, Titan might have local lakes or even seas of hydrocarbons.

Titan in the lab

The constituents of Titan's atmosphere are an oil company executive's dream. It may be that some Exxon spaceship would simply have to fly down, scoop up the stuff, and head home—or so the joke goes among those who realize that as our planet runs out of hydrocarbons, a virtually unlimited supply of them may rest as frozen or viscous pools of sludge on Titan. But exactly what are the constituents of this stuff?

When Carl Sagan late in life began to consider seriously the nature of Titan's atmosphere, he began a series of experiments attempting to replicate the chemicals. He was able to produce an organic sludge that was named tholin. Tholins are plastics, and this plastic may fall as rain from the dark Titan sky. In early 2004 chemists at the University of Arizona followed in Sagan's footsteps in re-creating Titan's atmosphere, but this time they used more accurate instrumentation and the vast store of new knowledge about Titan that had accumulated since Sagan's early and tragic death. They were surprised at the results. Titan is a Dow Chemical paradise.

In this new round of experiments, chemist Mark Smith and his colleagues bombarded a mixture of methane and nitrogen, the two major constituents of Titan's atmosphere, with high energy. Tholins are pretty gunky; this is the sort of stuff that is a cross between melted plastic and rancid butter. They do not dissolve in ethane, so on the surface of Titan there should be great gobs of this goo. But they do dissolve in water or ammonia, so if tholins can reach the under-ice ocean, or if a meteor crashes into Titan, or even if a volcano erupts, thus melting a large pool of water on the surface, the tholins will dissolve into amino acids, the very stuff of earth life. And if water appears, by whatever mechanism, it will not immediately refreeze since it is laced with ammonia, a wonderful antifreeze. With building blocks and energy, life has a chance—or had one. With each new asteroid strike on Titan the chemistry of life's formation is again put into play. Whether it gets to life will be answered only by our going there. I volunteer if Steve Benner will come with me.

So if the building blocks are there, where will Titan life find energy? This question has been examined by a variety of workers in a variety of ways. First, Bruce Jakosky and Everett Shock, more members of the NASA Astrobiology Institute, studied the heat flow on Mars in an effort to calculate how tectonically active Mars is. The method proved relevant to planets and moons other than Mars, including Titan. The two researchers found that there are two forms of heat coming from the interior of Titan. The first comes from radiogenic elements deep in the interior, just as it does on Earth. But unlike the earth, which has a huge quantity of heavy metals breaking down through radioactive fission processes, Titan has very little in the way of radioactive decay series. There is some heat coming from the deep interior, but not much compared with the earth or even Mars, which is very inactive compared with the earth. An equally important source of heat flow, it turns out, comes from an activity called tidal flexing. We have seen in chapter 11 that a similar source of energy heats the interior of Europa as well. Giant nearby Saturn has a profound gravity pull on Titan, and this attraction causes Titan's interior to flex, and heat, as it rotates around Saturn. These energy sources are probably sufficient to let life originate and maintain some sort of ecosystem but, compared with the earth, at only very low levels. Calculations suggest that the productivity of the Titan ocean (where productivity is a measure of the rate of carbon fixation from inorganic state to organic) is at least a thousand times lower than for the earth.

As important as energy is to life, so is nutrient availability. The open ocean of our Earth has more than enough energy from sunlight to maintain a rich biomass, yet the open oceans are virtual biological deserts. Energy is there to spare, but nutrients are not. For life as we know it (and, as we have seen, most life as we do not know it, some of which must certainly be carbon-based) the elements necessary are carbon, nitrogen, oxygen, hydrogen, phosphorus, sulfur, potassium, and sodium. A scarcity of any of these would severely hamper life's evolution and survival. Happily for any life trying to make or keep a foothold (cilia hold?) on Titan, each of these elements is probably present in the Titan oceans at concentrations comparable to earth seawater. Moreover, the oceans of Titan are probably strongly convective as a result of heat differences from bottom to top, and consequently the distribution of the listed elements should be rather uniform from top to the far distant bottom of the gigantic Titan subice oceans.

It is not just atoms in seawater that are necessary for life. Dissolved organic matter is a requirement for life in the sea on earth, where dissolved organic material varies between about one and seventy milligrams for every liter of seawater. There are two sources of this material in Titan seas: Chondritic asteroids crashing through the ice and penetrating into the sea, and from deep hydrothermal vents. Estimates allow that there are sufficient nutrients to maintain life in the Titan oceans.

What kind of earth life might survive on Titan?

What would a Titan bug eat? Although it has been estimated that there might be enough light to sustain photosynthesis in at least the shallowest parts of the Titan seas, a more likely scenario is for any life present to be chemolithotopic—as we have seen, a metabolism based on the breakdown of rock material, yielding hydrogen as an energy source. There are two such microbial metabolisms present on Earth that could possibly survive in the Titan ocean, based on what we listed earlier as the potential of energy and the presence of necessary elements for life. Both groups on Earth belong to the domain Archaea. One is the group known as methanogens; the other is denitrifying bacteria.

The methanogens group is of great interest to astrobiology. They appear to be very primitive, lowly members of the tree of life. They are obligate anaerobes. Oxygen will kill them, but on the early earth there was no oxygen, just as there is no oxygen on Titan. These little guys reduce carbon dioxide to produce methane. In this case, methane is a sign of life, and methanogens on Earth are very efficient at producing large volumes of this effective greenhouse gas. In the deep past the situation on Earth may have been even more efficient, effectively producing significant quantities of the atmosphere in this fashion. Is the methane found on Titan the consequence of life? This might be the most tantalizing question about Titan. A large population of methanogens near the surface of the Titan oceans could produce methane, which could make its way into the Titan atmosphere through cracks in the upper ice. The carbon supply for a population large enough to keep pumping methane into Titan's atmosphere would come from either planetary outgassing or the rain of meteorites.

Triton and Titan, a pair linked by panspermia?

Titan and Triton are the potential habitats of the most exotic life in the solar system, what I call silane life. Both are in regions where there are common and massive impacts by the wandering comets that surround our solar system, and there is the possibility that if one or the other evolved exotic silica-based life, the other would get it too. The delivery system would be the same that we have proposed for a potential passage of life from one place to another: panspermia processes.

Life in the atmosphere of the gas giants?

We should conclude this chapter with a long shot: the possibility of life in the atmosphere of the giant planets. Chapter 8 listed suggestions that there could conceivably be life in the atmosphere of Venus, and at least one biologist, Carl Woese, has proposed a means by which life itself might arise high in a planetary atmosphere. But in that latter case—for Earth—a great deal of molten iron would have been present, and such conditions would never be found in the gas giants at any time in their history. Jupiter, Saturn, Uranus, and Neptune are great gas globes with thick, roiling atmospheres. They are not unique to our solar system; in fact, they may be the most common class of planet that there is. The planet hunters are discovering new planets virtually every day now, in part because of their techniques but also because this class of planet might be very common in the universe. We know that gas giants like ours are common around other stars. At first glance these seem like rotten places for life to be. But this may not be the case. "Life finds a way" was a slogan in one of the *Jurassic Park* movies. Perhaps this is truer than we imagine. There is certainly a lot of real estate in the atmospheres of the gas giants. Could there be life there as well?

The late, great Carl Sagan certainly thought so. But then again, in a way he thought that there might be life everywhere. Sagan was a very good and clever scientist, and he (like Arthur C. Clarke) had visions of strange life floating in the clouds of Jupiter. Let's hope so. But there seems little in the way of science to support this idea.

The case for Titan

NASA is obsessed with Mars. Perhaps rightly so. It is the closest potentially habitable place to us (if we pass on the chance of Venusian atmospheric life) and thus would be the best chance of finding extraterrestrial aliens—or their fossils. But if we do find life there, it cannot be more than microbial, and even if it is CHON life that is shown to be unconnected to our own tree of life (I do not want to diminish the importance of this possibility), we would still, in all probability, be looking at a microbe that is equivalent to among the most primitive life on Earth: methanogens, or sulfate-reducing bacteria. As I discuss in the final chapter, the fossil record of Mars might be a much more exciting target altogether.

So if Mars offers hope at best for some pretty simple and perhaps familiar earthlike microbes, what about Titan? Titan, in my view, is another story.

Alone in the heavenly bodies of our solar system (besides Earth, of course, which has not yet been completely searched), Titan holds the promise of not just alien life but of more than one kind of fundamentally different alien life. Titan, in terms of life, could be home to three distinct empires of life, from two entirely different trees: CHON life of two kinds (ammono and water CHON life) and silicon life. The CHON ammono life would be found, presumably, beneath the ice, in the ammonia ocean, while earthlike CHON life would be found in the transient freshwater lakes on the Titan surface after an asteroid or comet impact. The silicon life would exist, if it existed at all, in the ethane-methane lakes of Titan's surface. No other place in our solar system would be so interesting to visit or hold out the promise of such exotic aliens. Our species, sooner or later, must visit Titan, to answer the most important biological question of all. Is there more than one way to make life naturally?

Back to Earth

With this chapter we finish our surveying voyage of the solar system. The prospects are exciting. The destinies of thousands of men and women now alive, or to be born, will be dedicated to understanding the diversity of life in the solar system. But let us come down to Earth, literally and figuratively, and ask some more difficult questions about aliens.

Chapter 13

Implications, Ethics, and Dangers

How the new disease spreads is known: it spreads in the cannibalism of
animals by animals, it spreads in the industrial cannibalism of animal
remains fed to animals, it spreads by the eating of beef.

—Richard Preston, *Deadly Feasts*

will start this chapter with a personal anecdote. In the earliest part
of the new century, some months after the publication of *Rare*
Earth by Don Brownlee and me, I was asked to appear on the PBS
television program *Evolution,* a program entirely funded by Seattle
native and resident Paul Allen, one of the cofounders of Microsoft
and thus one of the wealthiest men alive. My segment dealt with
mass extinction—namely, the single most catastrophic event in our
planet's history, the Permian extinction of 250 million years ago (an
extinction, by the way, that my research now shows to have been
caused by a sudden drop in atmospheric oxygen, not by the impact
of an asteroid on the earth). Since my work on this particular mass
extinction (this is my day job) had been carried out in South Africa's
Great Karoo Desert, I was flown there for the filming. It turned out
to be the first segment of the eight-hour series to be shot, and the
executive producer of the series, Richard Hutton, directed it himself.
We shot the segment, all went well, and about a year later I received

a phone call asking me to come to a science-planning session at the home of Paul Allen, at the behest of the same Richard Hutton, now working for Paul's Vulcan Inc.

The question that first came to mind was: *Which* home of Paul Allen's? This was answered on a rainy and windy Friday night when a large black limousine appeared at my house, drove me in the dark of late fall in Seattle, and deposited me in front of a seaplane. I was the only passenger. I had seen a science fiction movie that started out this way, *This Island Earth. Maybe Paul Allen is an alien* was the inevitable thought on the scary, hour-long trip through the buffeting black winds of a Gulf of Alaska–bred storm. And why not? The guy is scary smart. (Science is very lucky to have him funding so much good work. I am a huge admirer of the man and his sister Jody Patton.) After landing in the dark, I was whisked into another car and taken to a spectacular house on an island in the northern fringes of Puget Sound. Two long tables were filled with a who's who of American science, brought by Paul to spend a weekend imagining new ways of presenting science to the public. There were glittering (and really smart) people, food, and wine in the opulence of a book-lined great room and the storm howling futilely outside in the blackness, one of the great adventures of my life.

After an incredible dinner I made my way to our host and introduced myself. He was talking to a blond woman whom I did not know. It just so happened that I had a copy of my new book *Rare Earth,* already signed and as a present for Paul. By this time I had already participated in several public (radio and television) debates with the SETI honchos, and my point was that while the search for ET is an interesting venture, wouldn't it be better first to fix some of the disastrous catastrophes facing our planet, especially our biosphere with its rampant extinction of species, before spending huge quantities of cash on a radio telescope search for intelligent life? I had not considered the fact that this particular message was imperiling the livelihood of a bunch of people whose full-time jobs it was to look for aliens (and look for people to fund their look for aliens).

To them, I was the devil, threatening their angels, the rich technocrats with curiosity and money to burn.

I extended the book to Paul Allen, and to my surprise, his companion grimaced and shrank back at its sight. It was one of those movie moments when a cross is stuck in the face of a vampire (*It burns!*). Paul introduced me to his companion, who, it was clear, would rather that my plane had crashed and messily burned on takeoff or landing: thus I met Jill Tartar, head of SETI, the inspiration for the Ellie character in *Contact* by Carl Sagan. (Later, in 2004 she was named one of *Time* magazine's 100 Most Influential People in America; back then, in 2002, she was working up to the award.) At the time of this meeting she was trolling for Paul's money—twelve million dollars of it, in fact, to build what eventually was constructed and and named the Allen Telescope Array—to search for intelligent alien life. To her and SETI, *Rare Earth* was a principal threat, three hundred pages of reasons for people like Paul Allen not to fund SETI. I stood there, book in hand, as Ms. Tartar hustled Paul away.

In retrospect, I admit that I have been a little hard on SETI, for I too would love to *know* that there are other intelligences out there—and how will we find out but for SETI? Don Brownlee has convinced me that we humans will never make it to even the closest stars because of the vast distances and engineering obstacles. So to find out, we must search, and SETI is doing that about as efficiently as it can be done. But my feeling remains the same: There is no rush. So, is it better now to spend the money being used by SETI to save real life on Earth, as I advocated very publicly in 2001? In reality, the money spent on a year of SETI is less than we spend in a month (a day?) on foreign wars. The Iraq War alone could have sent several humans to Mars. So what about the NASA search for hypothetical life in the solar system, a search that costs billions, rather than the millions spent by SETI? Should we earmark *that* money for earthly environmental projects or health care for everyone in America? In this chapter I argue that we should continue on the path NASA has taken, studying the planets and moons of our solar system. In fact, here I will urge that we should expand that search by sending out two manned missions, one to Mars and one to the Saturnian moon Titan.

First, what would be the ramifications of SETI actually picking up a signal? Let us imagine that we wake up one morning, sleepily wander outside to search for wherever the morning paper has landed this time, and, coffee in hand, groggily look down at the blaring headline: SETI CONFIRMS SIGNAL FROM MU ARAE IS PROOF OF ALIEN INTELLIGENCE! Holy smoke! Aliens! Sorry, Jill and Seth! What happens next?

For a week it is huge news. Imagine the cable news stations:

"What in the world is mu Arae?"

"What a weird name!"

"Could this place be a source for new converts to Christianity/Islam/Buddhism/Scientology/etc.?"

"All credit goes to the Democrats/Republicans."

"Will Britney marry one?"

"Did Jesus save them too?"

"Can we sell them earth products?"

"Are they coming to conquer us?"

"Should we increase the defense budget?"

And on and on for the week. Turns out that mu Arae is the star orbited by the smallest extrasolar planet discovered to date, found in 2004. As extrasolar planets go, this one might be the first that is promising for CHON life. But this planetary system is fifty light-years from earth. The signal is nonspecific—it is not, "Hello, Earth, do you read us, over"—but it is definite proof of ET and, in a galaxy as large as ours, nearby ET at that. So a transmitter is rigged up, and out goes a signal, but what to send? No human language will do, so a bunch of prime numbers in sequence is flashed into space. And then we humans sit back and wait. Let's see, fifty years to get there, and if they receive it and then decide to send back a message, another fifty years go by. A century for a round-trip phone call. And what if they send back, "Message garbled, please repeat!"? No one alive will be there for the return message. Some conversation!

At any rate, now we know that we are not a unique intelligence in the universe. But who besides the religious ever doubted that? With more than four hundred billion stars in our Milky Way galaxy alone, how could we be unique? Rare, yes, unique, no. Pretty soon the talk shows and media personalities turn back to our problems. The never-ending war in Iraq. The never-ending global warming. The murder trial of the moment. Sports. Scandals. We have discovered that we are not alone, and in the end nothing has changed. So why would the discovery of microbial alien life in our solar system, or even on our planet, be any different?

The answer concerns knowledge and biology. There will be no wrong answer. We may spend billions and find that life is unique, at least in this solar system, to earth. This will be a powerful insight. The ramifications would be enormous. Every species going extinct on our planet as the result of our terrible stewardship and overpopulation would be another black mark against us.

What if we find only fossil remains of life on Mars and nothing on the other moons and planets? Again, a huge new insight: Life can form on (or be transported to) other worlds than Earth. And fossils are data: We would have some idea of what Martian life was like. How far in complexity did it advance? Did it get as far as multicellularity?

And what if there is life on Mars? What kind is it? Is it earthlike? What variety?

The greatest result would be life everywhere that we look. Some of it would have to be really alien. It has been said that just as the twentieth century was the century of physics, so our century will be that of biology. The discovery of life beyond Earth would be monumental. But are there caveats and dangers of our search? Certainly.

Planetary protection

In January 2004, NASA pulled off an enormous coup: It successfully landed the first of two rovers on Mars. The relief was palpable. Mars is littered with billions of dollars of failed missions, sad crash sites of smashed electronics and metal, representing heartbreak for those humans who spent years of their lives in mission design, then launch and journey, all the while with fingers crossed, worried but nevertheless believing that in the end *their* mission would not fall prey to that Bermuda Triangle of space, the area around Mars. Thus the jubilation on that early-winter morning of 2004.

I was jubilant too, of course, and bemused by another indication of how unfair life can be. The successful landing of the Mars rover *Spirit* was indeed a cause for celebration. But a much more audacious and difficult feat had been accomplished by NASA well beyond Mars the very day before, when the *Stardust* spacecraft had passed through Comet Wild 2, snatched some bits of the comet, and begun its return trip to Earth, with bits of the oldest particles in our solar system snugly packed away in its center. The Mars mission of the next day completely overshadowed a far more difficult feat, for as *Stardust* passed through the dust- and gravel-strewn region behind the giant comet, it ran through a field of fire, ran across the battlefield alive with a hail of bullets, and came out the other side unscathed. If it were to be done over ten times, would it have lived once more? Twice more? The odds of landing on Mars were far, far greater than successfully running through the comet tail as accomplished by *Stardust*.

The *Stardust* mission is something that I lived through vicariously by my day-to-day contact with its principal investigator, Don Brownlee, the most brilliant scientist that I know and one of the most enjoyable people on Planet Earth. *Stardust* blasted off in 1999, and at that time Don and I were in the midst of writing *Rare Earth*. Every day I traipsed up to his office and learned what it is like to have a satellite in space: the long telecoms with mission control, the almost weekly trips to Pasadena, the worry as glitch after glitch is overcome. In all this, I saw his cheerfulness depart him only once: In the second year of the mission, *Stardust*'s camera locked on the sun and was blinded, and the result was that for a day *Stardust* lost contact with Earth. That morning, when I made my daily trip to see Don, I found him pale, drawn, lifeless, and I knew that one of his "children" had been killed in the night. But the brilliance of a huge team on Earth saved Don's third child from certain death, and years later the same team skillfully veered *Stardust* into battle and through, with split-second decisions successfully made. Yet the next day NASA landed a rover on Mars, and *Stardust* was essentially forgotten. What has all this to do with the subject of my book? It is highly unlikely that microbes populated the tail or head of Comet Wild 2. Highly unlikely, but not impossible. And they are coming back to Earth. For all its brilliance and engineering magnificence, the *Stardust* mission represents something unique in human history. While we have brought back lots of rocks from the moon, this will be the first time that we will bring back pristine material from deep space that will not run through the sterilizing gauntlet of entry through the earth's atmosphere. *Stardust*'s precious cargo is scheduled to land in the Utah desert in 2006. NASA has deemed that there is no risk from the returning material, since Comet Wild 2 is frozen and, being small, is thought (at least by the bureaucrats) to be sterile of life. But Don Brownlee, while happy that he can get his material without waiting through a quarantine period, is not so sure that Wild 2 never had freshwater on it. What if it is a piece of a since-broken-apart planet like Pluto? We cannot say that there was never freshwater—and never life—on Wild 2. So it receives a "Get out of Jail Free" card. No planetary protection protocols at all. Alien material is coming to Utah and soon. Should we let it?

The hypothetical transfer of microbes from celestial body to body is panspermia, and as we saw in chapter 7, this has most likely been mediated (if it has happened at all) by the results of large-body impact. But a new form of panspermia has been happening since the start of the space age. While the various space agencies that have been flinging spacecraft onto various planetary surfaces have made concerted efforts to sterilize them prior to launch, it is an impossible task to send into space any spacecraft from earth that does not harbor any number of microbial stowaways. We have dropped our earth microbes on Venus, the moon, and Mars, and while the two former habitats have environments so extreme that they probably would not support life for very long, the microbes landing on Mars for almost decades now might have had a much better chance. This will be a great problem if we do find life on Mars. If it is identical to earth life, can we say that we did not deposit it there by our successes and failures on that planet?

There may be an even more egregious problem. The *Huygens* probe is now on the surface of Titan. Have we seeded Titan with early life? Again, how will we know when we do take samples of Titan and search them for life? We may have embarked on an interesting experiment: Titan may be able to support microbial life but not bring about its initial evolution. Are the earth microbes now there, even at this moment, starting their own peculiar line of evolution?

NASA is well aware of all these problems. This problem becomes more important over the next few years as the sample return missions from Mars begin to bring their payloads back to Earth. Like the *Stardust* mission, which is the vanguard of a new era of sample return science, we still have important decisions to make about safeguarding the earth from potential microbes from space. The decision has been made to allow outer space material to be brought back. How then can we protect ourselves from it if the tiny chance of its being inimical to earth life comes true?

Planetary protection is the term given to the policies and practices that protect other solar system bodies (e.g., planets, moons, asteroids, and comets) from earth life and that protect the earth from life that may be brought back from other solar system bodies. The question is: Which is in the greater danger, the earth from aliens or the other bodies in the solar system from us? This is not a new issue. The 1967 Treaty on Principles Governing the Activities of States in the Exploration and Use of Outer Space, Including the Moon and Other Bodies states that all countries party to the treaty "shall pursue studies of outer space, including the moon and other celestial bodies, and conduct exploration of them so as to avoid their harmful contamination." The technical aspects of planetary protection are determined by deliberations of the United Nations Committee on Space Research (COSPAR), part of the International Council of Science. The COSPAR Panel on Planetary Protection develops and makes recommendations on planetary protection policy. Since NASA is the most active agency in space these days, it is a key player in all this. NASA implements a planetary protection policy, and there is a planetary protection officer, who maintains overall planetary protection guidelines and assigns implementation requirements to each mission project. For years the NASA Planetary Protection Officer has been a great guy named John Rummel, and I admit to being a bit envious of his job. Imagine the cocktail conversation: "Hi, I'm John Rummel, planetary protection officer for Earth." Too bad he is an organization of one. Can you imagine how cool the uniforms would be in his Planetary Protection Force?

The most recent recommendations come from a report in 2000 that recommended that planetary missions have to be designed, cleaned, and operated in such a way as not to exceed a one in ten thousand chance of introducing any viable earth life to Mars or Europa. The spacecraft and its contents thus have to be *very* carefully cleaned and sometimes sterilized. After cleaning, the spacecraft is tested to assure that the cleanliness requirements have been met. The Planetary Protection Office can then certify that the requirements have indeed been met. In the case of *Stardust*, in his wisdom, Rummel and NASA are betting that Wild 2 is sterile, and they are probably right. But *probably* is not a comforting word to me.

So here they come, alien rocks from space that might contain alien life as well. But is an alien of any kind a danger?

Aliens from space as pathogens

The plot is a staple of science fiction, and why not? Alien microbes somehow fall to Earth, triggering a massive and lethal epidemic of some sort. The most famous twist on this was Michael Crichton's *Andromeda Strain.* So just how dangerous might alien life be to us? If we find and bring back Martian microbes, what are the chances that we will have opened Pandora's box or unleashed a new and deadly plague? The answer varies by type of alien, if this question can be answered at all. As a rough generalization, however, the more dissimilar the alien is to our form of biochemistry, the less dangerous it probably would be. A really alien biochemistry—our silicon life, for instance—would have little biological effect on us (unless, or course, its carbon side branches could somehow interfere with biochemical processes in DNA life that are crucial). In 1997 the National Research Council concluded that "if any organisms were returned from Mars that could survive on Earth, the potential for large scale ecological or pathogenic effects would be low."

The reality is that neither NASA nor anyone else really knows how dangerous the samples from a sample-return mission will be. There are clearly two risks: Alien bugs deleteriously affect our environments or are pathogenic and infectious in us or our agricultural animals and plants.

One of the scientists most interested in the problem, Andrew Schuerger, has considered the risk of aliens brought back from space to terrestrial plants and ecosystems. He defines a pathogen as a self-sustaining chemical system that derives its sustenance for a living cell or from the by-products of cell death. Disease, on the other hand, is the injurious alteration of one or more metabolic processes in the living host organism. The dangers of returned material from space is that it acts as an abiotic poison or as a biotic pathogen that secretes toxins dangerous to our life, or it directly invades terrestrial life as a disease agent. Schuerger comes at the problem of what we call planetary protection from the point of view of a plant pathologist. His take on the problem is reassuring: He thinks that there is very little danger from microbes brought back from space. Others are less certain. But aliens from space might not be the most dangerous aliens.

Engineered aliens as bioweapons

Probably of far greater danger than alien life from elsewhere in the solar system (or on our own planet) are alien life-forms created right on here on Earth specifically to serve as human-killing weapons—weapons of mass destruction, to be explicit. The National Academy of Sciences has looked long and hard at the development of new species of life specifically designed to kill humans, either directly or through the infection and death of human foodstuffs. The results of this study are frightening. It makes the following points:

- The effects of bioengineered life-forms could be worse than any disease known to humans.
- The broad wave of knowledge coming from the biotechnology revolution is expanding so rapidly and is so public that no current form of intelligence gathering can deal with the threat.

What is the specific threat from bioengineered life as we do not know it? Immense, it turns out. Just as we are struggling to make RNA in the lab, so others are constructing new viruses from scratch. One potential of such new life-forms is to cause infection and have the infected body—a human, for instance—start producing amino acids that are different from the twenty that we Terroans use. Already the bioweapons community has progressed to the point where smallpox viruses can be engineered to cause suppression of the human immune system during infection, making an already lethal disease very much more so. While these are not aliens per se, others are, including the designer biological weapons that have been engineered to cause our DNA to replicate falsely. These pathogens are aliens that take over our genomes and change them. Thus we see here something not mentioned previously, a life-form created to turn our own bodies into aliens by reacting in non-Terroan fashion to the infections agent. This is a distinction from using something such as smallpox to kill us, even a smallpox virus engineered to be more infectious than those in nature. An alien agent, engineered to be lethal, can cause our cells to do alien things.

Crop terrorism is another route. The average American city has food on hand for only five days. By attacking crops, bioweapons can wreck an economy quickly. Again, a simple way to do this is by producing an alien life-form that causes havoc in some key part of the complicated workings of DNA life.

What are the odds? Jack Szostak has estimated that it will take twenty million dollars to produce the first artificial life on Earth. Yet the budget for bioweapons research is in the billions of dollars per year in the United States alone. So chances are the next alien will not be one conceived to unlock the secrets of life but one engineered to create the dark world of unlife. Would those who killed thousands in the Twin Towers shrink from the chance to kill billions who have the temerity, like me, to be godless or who worship a different god?

Blessing and curse

What could be more exciting than the discovery of alien life? And what could be more frightening than the artificial creation of alien life? These are the two sides of the alien question. Which side do we choose?

Chapter 14

A Manifesto: Send Paleontologists to Mars and Biochemists to Titan

It was a new ship; it had fire in its belly, and men in its metal cells, and it moved with a clean silence, fiery and warm.

—Ray Bradbury, *The Martian Chronicles*

With all the ills facing our world, and the huge expense it will take to make *any* sort of dent in the effect that six billion humans have on our world as we all try to keep ourselves fed, healthy, and hydrated, can we afford space exploration? Can six billion humans coexist with nonhuman biodiversity at its current (and dwindling) number? Why would any rational scientist with a conscience advocate a series of hugely expensive, resource-squandering missions to far-off solar system real estate? Even a year ago I would certainly have protested such a proposal. But in this final chapter I argue for just this case: If we are ever to understand the frequency of life in the universe and the diversity of life that is possible, we need to send at least one manned mission to both Mars—which is certainly not a novel suggestion—and to Titan, which probably is. My rationale for this call will be the summary of the long journey that has been this book.

Send a paleontologist to Mars

More than three years ago my university decided that it wanted to hire two new faculty members in the field of astrobiology. There were already plenty of us around who by that time were calling ourselves astrobiologists, but here was an institutional commitment to the field. During the interview process of this search, I was struck by a comment made by one of the candidates, who, in fact, ultimately got one of these positions. His name is Roger Buick, and he is a paleontologist interested in the conditions and life on the earliest earth. Soon in our acquaintance, perhaps in a raucous and wine-filled evening (of which there are never enough with Roger, for in his cups his visions of the deep past come forth), he commented that "first off the initial manned spaceship landing on Mars should be a Buick." My initial reaction was "A car on Mars?" before I remembered who was regaling me. It turned out he was not joking. He wanted to be the first scientist to set foot on Mars, and for a good reason: Who better than a paleontologist to undertake the search for life on Mars?

He was right, of course. The first should be a paleontologist. It is no accident that paleontologists make up the largest number in the growing field of astrobiology, at least by initial discipline. Astrobiology is the prototypic multidisciplinary field, requiring deep expertise in more than one science, not just passing familiarity. Every paleontologist by necessity is a trained geologist and biologist as well. It requires the three fields to study life of the past. No shortcuts. For this reason I like to think that paleontologists are "preadapted" for the study of astrobiology. The NASA Astrobiology Institute certainly thinks so. If we look at the principal investigators of the institute, past and present, we see that there are more paleontologists running the individual nodes than any other discipline: Andrew Knoll of Harvard, Jack Farmer of Arizona State, Steven D'Hont from Rhode Island, Lisa Pratt from Indiana, Peter Ward from Washington, and the current director, Bruce Runnegar from UCLA, seals the deal.

So why send one to Mars? Because chances are the best we can hope for on Mars is a fossil. A paleontologist can be trained to take the necessary samples in the right places to look for extant microbes. It will not be hard to set up a drill rig to sample for Martian life in the permafrost. But a microbiologist could *never* be trained, on short notice, to find fossils. If I may go really fuzzy in these last pages, it is my belief that the best paleontologists come into the field because they are adept at making great fossil finds. Be it their eyes, understanding, instinct, certain people can be trusted to find fossils when it counts. Like a good RBI man in baseball, some are called to this, and can do so over and over. Some people just look harder, or longer, or simply want it more, and the fossils are found if they are there. We need someone like that first on Mars. Buick can do that.

Why might there be fossils? As we have seen in chapter 10, there is now irrefutable evidence that Mars long ago—in fact, at about the time that life first left evidence of itself on Earth (and as we have seen, *that* might be no accident)—was covered with at a minimum large lakes and perhaps true oceans. Such bodies of water leave behind sedimentary records. If there had been life, it will have left fossils of itself.

Microbes are soft and squishy and leave little record of themselves as individuals. But by their tendency to aggregate and multiply as hordes—on Earth at least—they have left undeniable records from the time of earliest earth life. The most visible of them are stromatolites, layered aggregates produced by the trapping of sediment by sheets of photosynthesizing bacteria. Microbes on Mars, if they evolved the mechanism of photosynthesis, probably would have used the same trick. As we have seen, the process of convergence is well known in evolution, and early life on Mars, even with some fundamentally different chemistry, would be discoverable.

What if something more complex than a bacterium emerged? In chapter 10 I showed how there might have been a rapid formation of an oxygenated atmosphere on Mars, allowing perhaps the evolution of something more complex than microbes. Could there have been the equivalent of a Cambrian explosion on Mars? Only the fossil record of Mars—if there is one—will tell us this, and it is a question that must be answered. If complex life evolved on Mars, Don Brownlee and I will have egg on our faces—but joyously so. Our Rare Earth Hypothesis posits lots of microbes and few planets where habitability stuck around long enough for complexity to arise. Perhaps complexity arises faster than we think, and Mars, with its early history being possibly quite earthlike, provides a test, but one based on fossils present or absent.

So why not let rovers sniff out any fossil record? Surely it would be cheaper and safer never to send humans to Mars. Can't a fossil search be conducted entirely by machine? To answer this question, let us conduct a thought experiment (one, unlike most thought experiments, that is entirely feasible). We should build a robot probe designed to look for fossils and test it by sending it to . . . Earth.

Our planet, with its unbelievably rich diversity, is without any doubt the richest oasis of life in our solar system and, if Don and I are correct, in many thousands of nearby solar systems. Yet if we think that life on our planet is currently diverse, with estimates of several to several tens of millions of extant species, how rich was the diversity of the entire past? How many species have there been in the history of life on our Earth? Hundreds of millions? Billions? My guess would be the latter. So with all that life, and so much of it leaving behind a skeletal record of itself in the fossil record, how difficult would it be to find fossils robotically?

The answer is probably that it would be hard to impossible to find fossils, even on Earth, with a machine. The reason is the spotty nature of the fossil record itself. Darwin, in later editions of his great book, railed against the fossil record and its inadequacies. He was not wrong. Even on this life-rich planet, fossil-bearing strata are rare. There is far more surface cover free of fossils than there is appropriate sedimentary rock that might yield fossils. If we landed our fossil-finding rover at random on land, chances are we would never find a fossil. Even limiting ourselves to sedimentary rock exposures where fossils might be present would be a fool's errand. That is why, earlier in this chapter, I touted the skill of some individuals in finding fossils; it is needed, even here. What chance would a rover have finding fossils on Mars if it could not be counted on to find them on Earth?

Thus my sense the that some young graduate student now undergoing the misery (joy, actually) of taking biology and geology courses might be the first off the lander. There is so much more, of course, but to me these are the most profound questions that Mars can answer: Did life evolve there, and how far in an evolution of complexity did it progress before a good planet went bad?

Send a biochemist to Titan

Mars is a traditional target, and it has been in NASA's sights for decades. Here I advocate a second and less traditional (and far more difficult) goal, a manned mission to Saturn's moon Titan. For this mission there will be no need for a paleontologist. First out of the Titan lander should be a biochemist. My colleague (and UW team member) Steve Benner should go, but I am afraid that he will be in the Great Biochemistry Laboratory in the Sky (or wherever biochemists go when they pass from chemical disequilibrium to equilibrium) before we ever get to Titan, if we as a species ever do. Come to think of it, Great Biochemistry Laboratory in the Sky might be a very relevant sobriquet for Titan, which indeed might be heaven for a biochemist even before he or she dies. One might even ask the ultimate teenage scientist question that we young paleontology grad students used to ask of each other: If you could go back in time and could thus really *know* the answers to so many unsolved questions, such as what dinosaurs *really* looked and lived like, yet could never come back, would you go? Back then we all avowed that we would go, even if it were a one-way trip. I suspect that not a few astrobiologists would take the same one-way trip if they could really find out if there is life on Titan and what *kind* of life it is. Unfortunately, chances are that any humans hazarding the long trip to the Saturnian system would be embarking on one-way trips. As dangerous as a mission to Mars would be, it pales in comparison with what would be required of the humans and machines leaving on the seven-or-more-year trip just to get from the earth to Titan. But actually to see the rings, to land on the ruddy moon, perhaps to make the most important biological discovery since Watson and Crick, I think that there would be no shortage of volunteers. Scores of terrorists blow themselves up yearly. Surely we can ask the same sacrifice for a better cause from our scientists, especially the older ones.

I am serious about beginning a drumbeat for a mission to Titan. Just as there are reasons that a very particular type of human needs to go to Mars, analogous arguments can be made about Titan. I have caught the excitement that Titan specialists such as Jonathan Lunine and Ralph Lorenz have long harbored about Titan, but my sense of its importance might be slightly different from theirs. To me, the most important aspect about Titan is not that it offers us a window of early biochemical systems present on Earth, *but that Titan is so different* from what is viewed as origin of life scenarios on Earth. That very alienness and the conditions discussed in chapter 12 make Titan an irresistible target for both biochemistry and biology. We could see how life might come about under conditions so different from those on Earth that a fundamentally different biology, if life is found, is almost assured. Titan might be the only place in the solar system with two radically different kinds of life—CHON life in its ammonia ocean, and ammono CHON life at that, and our one real hope to find the silanes: silicon life.

So there is my plea. Two different places, with two very different goals. Mars: to see if it was our early twin, where life came about. We could test the idea that given water and warmth, life is easily or readily made. That hypothesis would be sorely tested if we find Mars not only sterile now but also a place where life never came about, even with wet and warm. And Mars might help us with tests of panspermia.

And then Titan: to look for the most alien of life as we do not know it imaginable in the solar system. Titan, perhaps the most biologically diverse place in the solar system.

For all the reasons discussed in this book, let us start planning now. Our species can do this. Our species should do this. It may turn out, for other reasons that I can only vaguely suspect, that our species *must* do this. We may never come face-to-face with another intelligence. But there are so many reasons to try to come into the same room with life as we do not know it.

Envoi

A Forest of Life?

We have passed through 4.6 billion years of time and several billion miles in space to arrive here, these last pages. There has been fact, speculation, interpretation, and enthusiasm as we have looked at and classified earth and other life, smashed asteroids into things, built monsters, trashed movies, and otherwise run roughshod over life, alien and otherwise. Now I finish with some thoughts about a possible forest of evolutionary trees of life in the cosmos.

One of the great breakthroughs of the Western scientific tradition was the melding of evolutionary hypotheses with classificatory schemes. What a stroke of genius! With a simple diagram, we can both classify and portray an evolutionary pattern. Our tree diagrams both organize life on this planet and give the correct order of "begats."

In chapter 2 I made the bold move of suggesting that the hierarchical classification used to do this needs to have another formal level of category, one that I named a dominion. To finish the book and put it into context, I maintain that it is necessary to erect *yet another taxonomic category,* one that is above the dominions that were earlier defined. But how could there be anything higher than a dominion on our tree? There cannot, of course. As defined in these pages, a dominion is the highest possible category of taxonomic rank of the evolutionary tree of life on Earth. But what if we encounter life that has *absolutely* no connection to the evolutionary history of earth life? What if we discovered silane life, for instance?

The obvious answer to this, should we be lucky enough to witness its proof, is to create a new tree, one wholly independent of ours. A second tree of life. A tree of some alien life. We would have two trees known in the solar system. And suddenly, if all that we think and hope about the ease of making life is true, we have a new conception of the universe. Not as a starry firmament, or a background of low-temperature radiation, or a place of dark matter, or even some quirky place of strings and superstrings. To me, the universe is a forest. A forest of trees. A forest of trees of life of separate evolutionary creation, each tree unique in its own way. We sit on one tree in what I believe must be a vast and nearly infinite forest of trees, evolutionary trees, scattered through the galaxies of our universe. The trees themselves have to have a name. Let us call them Arborea.

Name of new taxonomic category: **Arborea**
Etymology: From the Latin, for tree
Definition. An Arborea is a taxonomic category that is above the level of dominion and is made up of all life coming from one independent genesis of life on a planet or moon. While it may spread to other planets, it has a single planet of origin.

Arborea Terra, Ward, 2005

Diagnosis: All CHON life evolved on Earth that is descended from one initial origin of life pathway.
Discussion: There will be a need for this particular taxonomic category only if there is indeed life beyond Earth or if we find or make life on Earth that has no biological connection in any way to earth life as we know it and it turns out that this life beyond earth life is not related to earth life. Perhaps there is but a single kind of life in the entire universe, spread far and wide by interstellar panspermia. In this case there would be but a single tree in the entire universe. But this seems highly unlikely, for reasons we have recounted in chapter 7. Far more likely is it that there will be as many, or nearly as many, trees in the universe as there are stars.

Are there other trees in our solar system? Is there an Arborea Ares, an Arborea Europa, an Arborea Titan? We shall know if we look. I hope we do. I hope that they are there. Loneliness is a terrible thing. It brings a responsibility that may be beyond our species and a responsibility to the great ark of Terroan life that we have dominion over. Surely there is more than one way to make life, as we have already shown in our laboratories. Are there aliens in existance? That question is already answered in the affirmative.

We are beginning to know life as we previously did not know it.

Afterword

I n the nearly three years since the original publication of this book, much has happened, both good and bad, with regard to the search for and synthesis of alien life. While no alien life has yet been found, efforts toward the artificial formation of life speed ahead. Also, new space probe data continues to come in, and, as always, there has been the unexpected development or two.

First, Harvard University made an enormous financial commitment to the work of Jack Szostak and his origin of life research. The recognition by Harvard, perhaps the preeminent science university on Earth, that the origin of life is one of the major scientific problems to be tackled indicates the degree of importance that the frontline scientific community attaches to origin of life studies. With the new money and personnel now attached to this project, we can expect a slew of new results in this and future decades.

Secondly, data from two space probes has added new information about our solar system and the possibilities of life elsewhere than Earth. The first of these was the successful return of the *Stardust* vehicle, which landed on target and on schedule in a Utah desert in early 2006. I stayed up that night, hoping for word as the principal investigator of this project was Don Brownlee, my friend and coauthor of *Rare Earth* and *The Life and Death of Planet Earth*. An hour before impact my phone rang: it was Don, from his cell phone, just calling in to break the nerve-racking wait at the remote landing site. Would the parachute open or would it, like its brother probe that had flown to the sun and back, smash into the Utah desert with its precious cargo, only to create a new crater in Utah?

Stardust behaved flawlessly; the parachute opened, and a helicopter rushed to the spot where the probe finally came to rest. Back at a nearby base, Don opened the probe and examined the collecting plate. Made of fine spun glass called Aerogel, bits of the comet Wild 5, whose head (not tail) *Stardust* had flown through, were indeed safely captured, and in great abundance.

The new discovery that amino acids apparently permeate the solar system as tiny comet particles is another great reason to believe that the stuff necessary to make life, Terroan-like or otherwise, is out there.

A second great discovery came from the Saturn-exploring *Cassini* probe. While its piggybacked *Huygens* probe was long since dead, *Cassini* eventually made its way back to the Titan region, and on a close flyby confirmed what the *Huygens* probe had suggested: that there are large pools of liquid on Titan's surface. It certainly is not water, and may be liquid hydrocarbon, even liquid methane, but it is liquid nevertheless. And as we have seen in the preceding pages, life is most probably liquid filled (rather than solid or gas), so this discovery only further enhances the case for going there someday in search of what might be the exotic life as we do not know it in the solar system.

Even Mars had its day of new evidence, although in this case it was not from the ever probing Mars *Rovers*, but from a hunk of Mars itself that fell to Earth and was collected early in the twentieth century.

In 1996, a team led by David McKay of NASA's Johnson Space Center published the sensational paper that elevated another piece of Mars meteorite to international fame and sparked one of the most remarkable scientific debates of recent times. As recounted earlier in these pages, they claimed that so-called ALH 84001 contained signatures of life, including small chainlike structures that could be fossil bacteria. If they were right, ALH 84001 contained the first aliens—in this case Martians—encountered by the human race.

The decade of study of ALH 84001 since has led to a verdict that the small and enigmatic structures found in ALH are not fossils, but were formed by inorganic processes that have nothing to do with life. Yet just as the original debate winds down, a new Martian meteorite is being quietly touted as evidence that life once existed on the red planet. This time it is not small rods, but tiny holes that have raised hopes—and ire—once again.

The meteorite in question is named Nakhla, which fell to Earth on June 28, 1911, at El-Nakhla in Egypt, reportedly killing a dog. The material arrived in around forty fragments, around ten kilograms in total. Later work on its chemistry indicated that it probably originated within thick lava flows that crystallized on Mars approximately 1.3 billion years ago, and was sent to Earth by a meteorite impact about 11 million years ago.

Unlike ALH, Nakhla shows that Mars had at least some water as late as 1.1 billion years ago, for it and the other nakhlites (Mars meteorites of similar chemical composition to Nakhla) have preserved some of the clearest traces of aqueous alteration within the parent rocks on Mars. In other words, minerals of Nakhla show clear signs that they were immersed in water sometime in their history. Water, that universal solvent and key to we Terroans, was certainly there in the past, as was independently found by the Mars *Rovers*. But these minerals are far different from the minerals of the sedimentary rocks that the *Rovers* photographed in such minute detail. It was in these new minerals that a team of astrobiologists found enigmatic but still controversial, evidence of past life there.

In 2006 a scientific team, led by Martin Fisk of Oregon State University in Corvallis, published a full report in the journal *Astrobiology* (Fisk et al. 2006 *Astrobiology* 6:48) of their new work and examinations of Nakhla. The evidence for possible life (at least past life) comes from tiny tubular holes that are commonly seen in volcanic rocks on Earth. These holes, which average 1 to 3 microns in diameter and up to 100 microns in length, are believed to be the results of bacteria or Archaea "eating" their way through newly formed basalts on the sea floor to obtain essential nutrients. Nobody has ever seen these bacteria in action, but there is evidence that the holes were formed this way. For example, the tunnels are exactly the right size to have been bored by bacteria, and often contain trace biomolecules such as nucleic acids. In fact, in 2004 geologists in Norway proposed that similar tunnels found in ancient basalts—around 3.5 billion years old—were the earliest signs of life on Earth.

To its surprise, Fisk's team found small tunnels virtually identical in size and shape to those in the terrestrial samples. Even more interesting, the tunnels were found in the same minerals observed to be tunneled on Earth. But unlike the tunnels in terrestrial rocks, Fisk and his crew were unable to extract any nucleic acids out of Nakhla's tunnels—though after experiencing liftoff and reentry, in addition to floating in space for 11 million years, the chance that this complex molecule could survive the journey in recognizable fashion in Nakhla's tunnels is pretty slim.

Other teams found carbonaceous matter in so called "veins" within Nakhla, and since the only kind of life that we know of (our own earth life) uses carbon, the presence of this element in abundance shows that the building blocks of life, in addition to water, were present when the rock that Nakhla came from formed. Could these enigmatic tunnels have originated from anything except life? So far, at least for similar minerals on Earth, the answer is no.

Not all are convinced by this new addition to the life on Mars parade. Joe Kirschvink of CalTech is one veteran of the ALH studies who, ironically enough (since he still believes that the tiny mineral chains on ALH were indeed made by life) is very skeptical that the tunnels on Nakhla are evidence that Mars once was the home to microbes that burrowed into rock in a manner similar to microbes on Earth. Kirschvink notes that the tunnels in the Nakhla minerals occur in roughly parallel orientation. While order is generally a sign of life, in this case it may be a sign of an abiotic origin for the tunnels, according to Kirschvink. All known tunnels made by Earth bacteria are far more random in orientation.

What, if anything, does this set of new observations on a Martian meteorite mean for further research on Mars? We now have a whole new kind of sign of life that could be detected by a robotic mission: the recognition of randomly oriented tubes, and if we are really lucky, tubes on Mars that have ancient nucleic acids within them. We even have information on where to look on Mars, and in rocks of a known age. A new NASA Mars mission scheduled for 2009 may send a robot with tools and targets informed in no small way by a dog-hitting rock from Mars.

These three examples of new work are not alone, of course. The very vibrancy of astrobiology as a field has resulted in a steady torrent of scientific papers, as well as material (such as this book) written for the interested public. It would thus be nice to conclude the book with recognition that all is well in the field. But such is not the case, and hence the bad news, very bad news indeed for astrobiologists. In early 2005, NASA announced that it was cutting funding to the field by half.

Luckily, the search for answers to so many questions about life is larger than even politics can control. The scientific searches will not be stopped, only delayed. We will make life, and I think and hope that we will find Life as We Do Not Know It.

References

Note: Unlike a scientific paper, in which each reference is documented after a statement, the reference list here refers to those papers relevant to and supporting the narrative of each chapter. While some references are used in more than a single chapter, each reference is listed only once, in the relevant chapter where it is first used or appears.

Preface and Introduction

Caldeira, K., and J. Kasting. 1992. The life span of the biosphere revisited. *Nature* 360:721–723.

————. 1992. Susceptibility of the early earth to irreversible glaciation caused by carbon ice clouds. *Nature* 359:226–28.

Dole, S. 1964. *Habitable Planets for Man.* Blaisdell Publishing Co: (New York).

Gott, J. 1993. Implications of the Copernican principle for our future prospects. *Nature* 363:315–19.

Gould, S. 1994. The evolution of life on earth. *Scientific American* 271:85–91.

Hart, M. 1979. Habitable zones around main sequence stars. *Icarus* 33:23–39.

Kasting, J. 1996. Habitable zones around stars: An update. In *Circumstellar Habitable Zones,* ed. L. Doyle. Travis House Publications: Menlo Park, Ca:) 17–28.

Kasting, J., D. Whitmire, and R. Reynolds. 1993. Habitable zones around main sequence stars. *Icarus* 101:108–28.

Laskar, J., and P. Robutel. 1993. The chaotic obliquity of planets. *Nature* 361:608–14.

Laskar, J.; F. Joutel; and P. Robutel. 1993. Stabilization of the earth's obliquity by the moon. *Nature* 361:615–17.

McKay, C. 1996. Time for intelligence on other planets. In *Circumstellar Habitable Zones,* ed. L. Doyle. Travis House Publications: (Menlo Park, Ca.) 405–19.

Schwartzman, D., and S. Shore. 1996. Biotically mediated surface cooling and habitability for complex life. In *Circumstellar Habitable Zones,* ed. L. Doyle. Travis House Publications: 421–43.

Volk, T. 1998. *Gaia's Body: Toward a Physiology of Earth.* Copernicus: (Newton, Mass.).

Walker, J.; P. Hays; and J. Kasting. 1981. A negative feedback mechanism for the long-term stabilization of earth's surface temperature. *JGR* 86:9776–82.

Ward, P., and D. Brownlee. 2001. *Rare Earth: Why Complex Life Is Uncommon in the Universe.* Copernicus Press: (Newton, Mass.).

Chapter 1. What Is Life?

Cronier, S., et al. 2004. Prions can infect primary cultured neurons and astrocytes and promote neuronal cell death. *PNAS* 101 (33):12271–276.

Davies, Paul. 1998. *The Fifth Miracle.* Alan Lane/Penguin Press: (London).

Dyson, F. 1999. *Origins of Life,* 2nd ed. Cambridge University Press: (New York).

Follet, J., et al. 2002. PrP expression and replication by Schwann cells: Implications in prion spreading. *J. Virol.* 76(5):2434–39.

Haldane, J. 1947. *What Is Life?* Boni and Gaer: (New York).

Jackson, G. S., and J. Collinge. 2001. The molecular pathology of CJD: Old and new variants. *J. Clin. Pathol.: Mol. Pathol.* 54:393–99.

Legname, G., et al. 2004. Synthetic mammalian prions. *Science* 305:673–76.

Luisi, P. L. 1998. About various definitions of life. *Origins Life Evolution Biosphere* 28:613–22.

Olomucki, M. 1993. *The Chemistry of Life.* McGraw-Hill, Inc.: (New York).

Orgel, L. 1973. *The Origins of Life: Molecules and Natural Selection.* John Wiley and Sons: (Hoboken, N.J.).

Schrödinger, E. 1944. *What Is Life?* Cambridge University Press: (New York).

Stetter, K. O. 2002. Hyperthermophilic microorganisms. In *Astrobiology: The Quest for the Conditions of Life,* ed. G. Horneck and C. Baumstark-Khan. Springer: (New York) 169–84.

Chapter 2. What Is Earth Life?

Bada, J. L.; C. Bigham; and S. L. Miller. 1994. Impact melting of frozen oceans on the early earth: Implications for the origin of life. *Proceedings National Academy of Science USA* 91:248–1250.

Barns, S. M.; R. E. Fundyga; M. W. Jeffries; and N. R. Pace. 1994. Remarkable archaeal diversity detected in a Yellowstone National Park hot spring environment. *Proc. Nat'l. Acad. Sci. USA* 91:1609–13.

Baross, J. A., and J. W. Deming. 1995. Growth at high temperatures: Isolation and taxonomy, physiology, and ecology. In *The Microbiology of Deep-Sea Hydrothermal Vents,* ed. D. M. Karl, CRC Press: (Boca Raton, Fla.) 169–217.

Baross, J. A., and S. E. Hoffman. 1985. Submarine hydrothermal vents and associated gradient environments as sites for the origin and evolution of life. *Orig. Life Evolution Biosphere* 15:327–45.

Converse, D. R.; H. D. Holland; and J. M. Edmond. 1984. Flow rates in the axial hot springs of the East Pacific Rise (21°N): Implications for the heat budget and the formation of massive sulfide deposits. *Earth Planet. Sci. Lett.* 69:159–75.

Daniel, R. M. 1992. Modern life at high temperatures. In *Marine Hydrothermal Systems and the Origin of Life,* ed. N. Holm. Special issue of *Orig. Life Evolution Biosphere* 22:33–42.

Giovannoni, S. J.; T. D. Mullins; and K. G. Field. 1995. Microbial diversity in oceanic systems: rRNA approaches to the study of unculturable microbes. *Molecular Ecology of Aquatic Microbes,* ed. I. Joint. Springer-Verlag: (New York) 217–48.

Glikson, A. 1995. Asteroid comet mega-impacts may have triggered major episodes of crustal evolution. *Eos* 76:49–54.

Helgeson, H. C.; A. M. Knox; C. E. Owens; and E. L. Shock. 1993. Petroleum, oil field waters and authigenic mineral assemblages: Are they in metastable equilibrium in hydrocarbon reservoirs? *Geochim. Cosmochim Acta,* 57:3295–3339.

Karl, D. M. 1995. Ecology of free-living, hydrothermal vent microbial communities. *The Microbiology of Deep-sea Hydrothermal Vents,* ed. D. M. Karl. CRC Press: (Boca Raton, Fla.) 35–124.

Miller, S., and A. Lazcano. 1996. From the primitive soup to Cyanobactyeria: It may have taken less than 10 million years. In *Circumstellar Habitable Zones* ed. L. Doyle. Travis House Publications: (Menlo Park, Ca.) 393–404.

Norton, D. L. 1984. Theory of hydrothermal systems. *Ann. Rev. Earth Planet. Sci.,* 12:155–77.

Pace, N. R. 1991. Origin of life: Facing up to the physical setting. *Cell* 65:531–33.

Russell, M. J.; A. J. Hall; A. G. Cairns-Smith; and P. S. Braterman. 1988. Submarine hot springs and the origin of life. *Nature* 336:117.

Russell, M. J.; A. J. Hall; and D. Turner. 1989. In vitro growth of iron sulphide chimneys: Possible culture chambers for origin-of-life experiments. *Terra Nova* 1:238–41.

Russell, M. J.; R. M. Daniel; and A. J. Hall. 1993. On the emergence of life via catalytic iron sulphide membranes. *Terra Nova* 5:343–347.

Schwartzman, D.; M. McMenamin; and T. Volk. 1993. Did surface temperatures constrain microbial evolution? *BioScience* 43:390–93.

Segerer, A. H.; S. Burggraf; G. Fiala; G. Huber; R. Huber; U. Pley; and K. O. Stetter. 1993. Life in hot springs and hydrothermal vents. *Orig. Life Evol. Biosphere* 23:77–90.

Shock, E. L. 1990. Geochemical constraints on the origin of organic compounds in hydrothermal systems. *Orig. Life Evol. Biosphere* 20:331–67.

———. 1992. Stability of peptides in high temperature aqueous solutions. *Geochim. Cosmochim. Acta* 56:3481–91.

Shock, E. L.; T. McCollom; and M. D. Schulte. 1995. Geochemical constraints on chemolithoautotrophic reactions in hydrothermal systems. *Orig. Life Evol. Biosphere* 25:141–59.

Sleep, N. H.; K. J. Zahnle; J. F. Kasting; and H. J. Morowitz. 1989. Annihilation of ecosystems by large asteroid impacts on the earth. *Nature* 342:139–42.

Stetter, K. O. 1995. Microbial life in hyperthermal environments. *ASM News* 61:285–90.

Takai, K.; T. Komatsu; F. Inagaki; and K. Horikoshi. 2001. Distribution of archaea in a black smoker chimney structure. *Appl. Environ. Microbiol.* 67:3618–29.

Thomas, D. N., and G. S. Dieckmann. 2002. Antarctic sea ice: A habitat for extremophiles. *Science* 295:641–44.

Von Damm, K. L. 1990. Seafloor hydrothermal activity: Black smoker chemistry and chimneys. *Ann. Rev. Earth Planet. Sci.:* 173–204.

Wilson, E. 1992. *The Diversity of Life.* Harvard University Press: (Cambridge, Mass.).

Woese, C. R. 1987. Bacterial evolution. *Microbiol. Rev.* 51:221–71.

Woese, C. R.; O. Kandler; and M. L. Wheelis. 1990. Towards a natural system of organisms: Proposal for the domains Archaea, Bacteria, and Eucarya, *Proc. Nat'l. Acad. Sci. USA* 87:4576–79.

Chapter 3. Life As We Do Not Know It

Bains, W. 2001. The parts list of life. *Nat. Biotechnol.* 19:401–02.

Baines, W. 2004. Many chemistries could be used to build living systems. *Astrobiology* 4:137–67.

Brook, M. A. 2000. *Silicon in Organic, Organometallic and Polymer Chemistry.* John Wiley and Sons: (Hoboken, N.J.).

Cairns-Smith, Alexander. 1982. *Genetic Takeover and the Mineral Origins of Life.* Cambridge University Press: (New York).

Crawford, R. L., et al. 2001. In search of the molecules of life. *Icarus* 154:531–39.

Dabrowska, B. 1984. The solubility of selected halogen hydrocarbons in liquid nitrogen at $77.4°$ K. *Cryogenics* 24:276–77.

Feinberg, G., and R. Shapiro. 1980. *Life Beyond Earth.* William Morrow: (New York).

Firsoff, V. 1962. An ammonia based life. *Discovery* 23:36–42.

———. 1965. Possible alternative chemistries of life. *Spaceflight* 7:132–36.

Franck, S.; W. von Bloh; C. Bounama; M. Steffen; D. Schonberner; and H. J. Schellnhuber. 2000. Determination of habitable zones in extrasolar planetary systems: Where are Gaia's sisters? *J. Geophys. Res.* 105:1651–58.

Grinspoon, D. 2004. *Lonely Planets.* HarperCollins: (New York).

Harrison, P. G. 1997. Silicate cages: Precursors to new materials. *J. Organometal. Chem.* 542:141–84.

Haldane, J. 1954. The origins of life. *New Biology* 16:12–27.

Irwin, L. N., and D. Schulze-Makuch. 2001. Assessing the plausibility of life on other worlds. *Astrobiology* 1:143–60.

Koerner, D., and S. LeVay, 2000. *Here Be Dragons.* Oxford University Press: (New York).

Lickiss, P. 2001. Polysilanols. In *The Chemistry of Organic Silicon Compounds,* vol. 3, ed. Z. Rappoport and Y. Apeloig. John Wiley and Sons: (Hoboken, N.J.) 695–744.

Lovelock, J. 1979, *Gaia, a New Look at Life on Earth.* Oxford University Press: (New York).

McCarthy, M. C.; C. A. Gottlieb; and P. Thaddeus. 2003. Silicon molecules in space and in the laboratory. *Mol. Phys.* 101:697–704.

McCollom, T. M., and J. S. Seewald. 2001. A reassessment of the potential for reduction of dissolved CO_2 to hydrocarbons during serpentization of olivine. *Geochim. Geophys. Acta* 65:3769–78.

Pace, N. R. 2001. The universal nature of biochemistry. *Proc. Natl. Acad. Sci. USA* 98:805–08.

Patai, S., and Z. Rappoport, eds. 1989. *The Chemistry of Organic Silicon Compounds,* John Wiley and Sons: (Hoboken, N.J.).

Saenger, W. 1987. Structure and dynamics of water surrounding biomolecules. *Annu. Rev. Biophys. Biophys. Chem.* 16:93–114.

Sagan, C. 1973. Extraterrestrial life. In *Communication with Extraterrestrial Intelligence (CETI)*, ed. C. Sagan. MIT Press: (Cambridge, Mass.). 42–67.

Stroppolo, M. E.; M. Falconi; A. M. Caccuri; and A. Desiderim. 2001. Superefficient enzymes. *Cell. Mol. LifeSci.* 58:1451–60.

Tacke, R., and H. Linoh. 1989. Bioorganosilicon chemistry. In *The Chemistry of Organic Silicon Compounds,* Part 2, ed. S. Patia and Z. Rappoport. John Wiley and Sons: (Hoboken, N.J.) 1143–1206.

West, R. 2002. Multiple bonds to silicon: 20 years later. *Polyhedron* 21:467–72.

Whitesides, G. M., and Grzybowski, B. 2002. Self-assembly at all scales. *Science* 295:2418–21.

Chapter 4. A Recipe Book of Life

Bachmann, P. A.; P. L. Luisi; and J. Lang. 1992. Autocatalytic self-replication micelles as modes for prebiotic structures. *Science* 357:57–58.

Baross, J. A., and S. E. Hoffman. 1985. Submarine hydrothermal vents and associated gradient environments as sites for the origin and evolution of life. *Origins of Life* 15:327–45.

Baross, J. A., and J. F. Holden. 1996. Overview of hyperthermophiles and their heat-shock proteins. *Advances in Protein Chemistry* 48:1–35.

Benner, S. A. 2004 (in press). Understanding nucleic acids using synthetic chemistry. *Acc. Chem. Res.*

Benner, S. A., and D. Hutter. 2002. Phosphates, DNA, and the search for nonterran life: A second generation model for genetic molecules. *Bioorg. Chem.* 30:62–80.

Breslow, R. 1959. On the mechanism of the formose reaction. *Tetrahedron Lett.* 21:22–26.

Bernal, J. D. 1951. *The Physical Basis of Life.* Routledge and Kegan Paul: (London).
———. 1967. *The Origin of Life.* World: (Cleveland, Oh.).

Cairns-Smith, A. G. 1982. *Genetic Takeover and the Mineral Origins of Life.* Cambridge University Press: (New York).

Carter, C.W., Jr., and J. Kraut. 1974. A proposed model for interaction of polypeptides with RNA. *Proc. Nat. Acad. Sci. USA* 71:283–87.

Cech, T. R. 1986. A model for the RNA-catalyzed replication of RNA. *Proc. Nat. Acad. Sci. USA* 83:4360–63.

————. 1987. The chemistry of self-splicing RNA and RNA enzymes. *Science* 236:1532–39.

Cech, T. R., and B. L. Bass. 1986. Biological catalysis by RNA. *Annual Review of Biochemistry* 55:599–629.

Chang, S. 1994. The planetary setting of prebiotic evolution. In *Early Life on Earth*, Nobel Symposium No. 84, ed. S. Bengston. Columbia University Press: (New York) 10–23.

Darwin, C. R. 1859. *On the Origin of Species by Means of Natural Selection or the Preservation of Favored Races in the Struggle for Life.* John Murray: (London).

Dawkins, R. 1976. *The Selfish Gene.* Oxford University Press: (New York).

Deamer, D. W. 1992. Polycyclic aromatic hydrocarbons: Primitive pigment systems in the prebiotic environment. *Adv. Space Res.* 12:183–89.

Deamer, D. W., and E. Harang. 1990. Light-dependent pH gradients are generated in liposomes containing ferrocyanide. *BioSystems* 24:1–4.

Deamer, D. W., and R. M. Pashley. 1989. Amphiphilic components of the Murchison carbonaceious chondrite: Surface properties and membrane formation. *Origin of Life and Evolution of the Bioshpere* 19:21–38.

Doolittle, W. F., and J. R. Brown. 1994. Tempo, mode, the progenote, and the universal root. *Proceedings National Academy of Sciences USA,* 91:6721–28.

Doolittle, W. F.; D. F. Feng; S. Tsang; G. Cho; and E. Little. 1996. Determining divergence times of the major kingdoms of living organisms with a protein clock. *Science* 271:470–77.

Doudna, J. A., and J. W. Szostak. 1989. RNA-catalysed synthesis of complementary-strand RNA. *Nature* 339:519–22.

Eigen, M. 1992. *Steps Towards Life: A Perspective on Evolution.* Oxford University Press: (New York).

Forterre, P. 1997. Protein versus rRNA: Problems in rooting the universal tree of life. *American Society for Microbiology News,* 63, 89–95.

Forterre, P.; F. Confalonieri; F. Charbonnier; and M. Duguet. 1995. Speculations on the origin of life and thermophily: Review of available information on reverse gyrase suggests that hyperthermophilic procaryotes are not so primitive. *Origins of Life and Evolution of the Biosphere* 25:235–249.

Fox, S.W. 1965. Simulated natural experiments in spontaneous organization of morphological units from protenoid. In *The Origins of Prebiological Systems and Their Molecular Matrices,* ed. S.W. Fox. Academic Press: (New York) 361–82.

————. 1980. The origins of behavior in macromolecules and protocells. *Comparative Biochemistry and Physiology* 67B:423–36.

————. 1988. *The Emergence of Life: Darwinian Evolution from the Inside.* Basic Books: (New York).

————. 1995. Thermal synthesis of amino acids and the origin of life. *Geochemica et Cosmochimica Acta* 59:1213–14.

Gesteland, R. F., and J. F. Atkins. 1993. *The RNA World: The Nature of Modern RNA*

Suggests a Prebiotic RNA World. Cold Spring Harbor Laboratory Press: (Stony Brook, N.Y.).

Gogarten-Boekels, M.; E. Hilario; and J. P. Gogarten. 1995. *Origins of Life and Evolution of the Biosphere,* 25:251–64.

Grayling, R. A.; K. Sandman; and J. N. Reeve. 1996. DNA stability and DNA binding proteins. *Advances in Protein Chemistry* 48:437–67.

Gu, X. 1997. The age of the common ancestor of eukaryotes and prokaryotes: Statistical inferences. *Molecular Biology and Evolution* 14:861–66.

Gupta, R. S., and G. B. Golding. 1996. The origin of the eukaryotic cell. *Trends in Biochemical Sciences,* 21:166–71.

Haldane, J. B. S. 1929. The origin of life. Rationalist animal. Reprinted in *The Origin of Life,* ed. J. D. Bernal. 1967. World: 242–49.

Harold, F. M. 1986. *The Vital Force: A Study of Bioenergetics.* Freeman: (New York).

Hedén, C-G. 1964. Effects of hydrostatic pressure on microbial systems. *Bacteriological Reviews* 28:14–29.

Hei, D. J., and D. S. Clark. 1994. Pressure stabilization of proteins from extreme thermophiles. *Applied and Environmental Microbiology* 60:932–39.

Hennet, R. J.-C.; N. G. Holm; and M. H. Engel. 1992. Abiotic synthesis of amino acids under hydrothermal conditions and the origin of life: A perpetual phenomenon? *Naturwissenschaften* 79:361–65.

Hilario, E., and J. P. Gogarten. 1993. Horizontal transfer of ATPase genes: The tree of life becomes the net of life. *BioSystems* 31:111–19.

Holden, J. F., and J. A. Baross. 1995. Enhanced thermotolerance by hydrostatic pressure in deep-sea marine hyperthermophile *Pyrococcus* strain ES4. *FEMS Microbiology Ecology* 18:27–34.

Holden, J. F.; M. Summit; and J. A. Baross. 1997. Thermophilic and hyperthermophilic microorganisms in 3–30°C hydrothermal fluids following a deep-sea volcanic eruption. *FEMS Microbiology Ecology* (in press).

Huber, R.; P. Stoffers; S. Hohenhaus; R. Rachel; S. Burggraf; H. W. Jannasch; and K. O. Stetter. 1990. Hyperthermophilic archaebacteria within the crater and open-sea plume of erupting MacDonald Seamount. *Nature* 345:179–82.

Hunten, D. M. 1993. Atmospheric evolution of the terrestrial planets. *Science* 259:915–20.

Joyce, G. F. 1994. Introduction. In *Origins of Life: The Central Concepts,* eds. D. W. Deamer and G. R. Fleischaker. Jones & Bartlett: (Boston, Mass.) xi–xii.

Joyce, G. F., and L. E. Orgel. 1993. Prospects for understanding the origin of the RNA world. In *The RNA World,* eds. R. F. Gesteland and J. F. Atkins. Cold Spring Harbor Laboratory Press: (Stony Brook, N.Y.) 1–25.

———. 1999. Prospects for understanding the origin of the RNA world. In *The RNA World,* 2nd ed., eds. R. Gestland, J. Atkins, and T. Cech. Cold Spring Harbor Press: (Stony Brook, N.Y.).

Kadko, D.; J. Baross; and J. Alt. 1995. The magnitude and global implications of hydrothermal flux. In *Physical, Chemical, Biological and Geological Interactions Within Sea Floor Hydrothermal Discharge,* Geophysical Monograph 91, eds. S. Humphris, R. Zierenberg, L. Mullineaux, and R. Thompson. AGU Press: (Washington, D.C.) 446–66.

Kauffman, S. A. 1993. *The Origins of Order: Self-Organization and Selection in Evolution.* Oxford University Press: (New York).

Knauth, L. P. 1998. Salinity history of the earth's early ocean. *Nature* 395:554–55.

Knauth, L. P., and S. Epstein. 1976. Hydrogen and oxygen isotope ratios in nodular and bedded cherts. *Geochimica et Cosmochimica Acta* 40:1095–1108.

Krüppers, B. O. 1990. *Information and the Origin of Life.* MIT Press: (Cambridge, Mass.).

Langton, C. G. 1986. Studying artificial life with cellular automata. *Physica* (D) 22:120–49.

———. 1992. Life at the edge of chaos. In *Artificial Life* II, eds. C. G. Langton, C. Taylor, J. D. Farmer, and S. Rasmussen. Addison-Wesley: (Boston, Mass.) 41–91.

Larralde, R.; M. P. Robertson; and S. L. Miller. 1995. Rates of decomposition of ribose and other sugars: Implications for chemical evolution. *Proc. Natl. Acad. Sci. USA* 92:8158–60.

Lazcano, A. 1994. The RNA world, its predecessors, and its descendants. *In Early Life on Earth,* ed. S. Bengston. Columbia University Press: (New York) 70–80.

Lorsch, J. R., and J. W. Szostak. 1994. In vitro evolution of new ribozymes with polynucleotide kinase activity. *Nature* 371:31–36.

Lowe, D. R. 1994. Early environments: Constraints and opportunities for early evolution. In *Early Life on Earth,* Nobel Symposium No. 84, ed. S. Bengston. Columbia University Press: (New York) 24–35.

Maher, K. A., and J. D. Stevenson. 1988. Impact frustration of the origin of life. *Nature* 331:612–14.

Marshall, W. L. 1994. Hydrothermal synthesis of amino acids. *Geochimica et Cosmochimica Acta* 58: 2099–2106.

Miller, S. L. 1953. A production of amino acids under possible primitive earth conditions. *Science* 117:528–29.

Miller, S. L., and J. L. Bada. 1988. Submarine hot springs and the origin of life. *Nature* 334:609–11.

Mitchell, P. D. 1979. Keilin's respiratory chain concept and its chemiosmotic consequences. *Science* 206:1148–59.

Mojzsis, S.; G. Arrhenius; K. D. McKeegan; T. M. Harrison; A. P. Nutman; and C. R. L. Friend. 1966. Evidence for life on earth before 3,800 million years ago. *Nature* 385:55–59.

Moorbath, S.; R. K. O'Nions; and R. J. Pankhurst. 1973. Early Archaean age of the Isua iron formation. *Nature* 245:138–39.

Morowitz, H. 1992. *Beginnings of Cellular Life: Metabolism Recapitulates Biogenesis.* Yale University Press: (Cambridge, Mass.).

Morowitz, H., D. W. Deamer; and T. Smith. 1991. Biogenesis as an evolutionary process. *Journal of Molecular Evolution* 33:207–08.

Oparin, A. I. 1924. *Proiskhozhdenie zhizy.* Moskovski Rabochii.

————. 1938. *The Origin of Life.* Macmillan: (New York).

Orgel, L. E. 1968. Evolution of the genetic apparatus. *Journal of Molecular Biology* 38:381–93.

————. 1992. Molecular replication. *Nature* 358:203–09.

Ourisson, G., and Y. Nakatani, 1994. The terpenoid theory of the origin of cellular life: The evolution of terpenoids to cholesterol. *Current Biology* 1:11–23.

Pace, N. 1991. Origin of life: Facing up to the physical setting. *Cell* 65:531–33.

Pedersen, K. 1993. The deep subterranean biosphere. *Earth Sci. Rev.* 34:243–60.

Ricardo, A.; M. A. Carrigan; A. N. Olcott; and S. A. Benner. 2004. Borate minerals stabilize ribose. *Science* 303:196.

Russell, M. J., and A. J. Hall. 1997. The emergence of life from iron monosulphide bubbles at a submarine hydrothermal redox and pH front. *J. Geol. Soc. Lond.* 154:377–402.

Sagan, C., and C. Chyba. 1997. The early faint sun paradox: Organic shielding of ultraviolet-labile greenhouse gases. *Science* 276:1217–21.

Salthe, S. N. 1991. Formal consideration on the origin of life. *Uroboros* 1:45–65.

Schopf, J. W. 1994. The oldest known records of life: Early archean stromatolites, microfossils, and organic matter. In *Early Life on Earth,* ed. S. Bengston. Columbia University Press: (New York) 193–206.

Schopf, J. W., and B. M. Packer. 1987. Early archean (3.3-billion- to 3.5-billion-year-old) microorganisms from the Warrawoona Group, Australia. *Science* 237:70–73.

Shock, E. L. 1992. Chemical environments of submarine hydrothermal systems. *Origin of Life and Evolution of the Biosphere* 22:67–107.

Sogin, M. L. 1991. Early evolution and the origin of eukaryotes. *Current Opinion in Genetics and Development* 1:457–63.

Sogin, M. L.; J. D. Silverman; G. Hinkle; and H. G. Morrison. 1996. Problems with molecular diversity in the Eucarya. In *Society for General Microbiology Symposium: Evolution of Microbial Life,* eds. D. M. Roberts, P. Sharp, G. Alderson, and M. A. Collins. Cambridge University Press: (New York) 167–84.

Stetter, K. O.; R. Huber; E. Blöchl; M. Kurr; R. D. Eden; M. Fielder; H. Cash; and I. Vance. 1993. Hyperthermophilic archaea are thriving in deep North Sea and Alaskan oil reservoirs. *Nature* 365:743–45.

Stevens, T. 1997. Lithoauthtotrophy in the subsurface. *FEMS Microbiol. Rev.* 20:327–37.

Swenson, R. 1989. Emergent attractors and the law of maximum entropy production: Foundation to a theory of general evolution. *Systems Research* 6:187–97.

————. 1998. *Spontaneous Order, Evolution and Natural Law: An Introduction to the Physical Basis for an Ecological Psychology.* Lawrence Erlbaum Associates: (Mahwah, N.J.).

Wang, J.-F.; W. D. Downs; and T. R. Cech. 1993. Movement of the guide sequence during RNA catalysis by a group I ribozyme. *Science* 260:504–08.

Weber, B. H.; D. J. Depew; C. Dyke; S. N. Salthe; E. D. Schneider; R. E. Ulanowicz;

and J. S. Wicken. 1989. Evolution in thermodynamic perspective: An ecological approach. *Biology and Philosophy* 4:373–405.

Weber, B. H., and D. J. Depew. 1996. Natural selection and self-organization: Dynamical models as clues to a new evolutionary synthesis. *Biology and Philosophy* 11:33–65.

Westheimer, F. H. 1987. Why nature chose phosphates. *Science* 235:1173–78.

Wicken, J. S. 1987. *Evolution, Information and Thermodynamics: Extending the Darwinian Program.* Oxford University Press: (New York).

Woese, C. 1967. The evolution of the genetic code. In *The Genetic Code.* Harper and Row: (New York) 179–95.

———. 1994. There must be a prokaryote somewhere: Microbiology's search for itself. *Microbiological Reviews* 58:1–9.

Woese, C.; O. Kandler; and M. L. Wheelis. 1990. Towards a natural system of organisms: Proposals for the domains Archaea, Bacteria, and Eucarya. *Proceedings National Academy of Sciences USA* 87:4576–79.

Chapter 5. The Artificial Synthesis of Life

Deamer, D. W., and R. M. Pashley. 1989. Amphiphilic components of the Murchison carbonaceous chondrite: Surface properties and membrane formation. *Origins Life Evol. Biosphere* 19:21–38.

Duve, D. C. 1998. Clues from present-day biology: The thioester world. In *The Molecular Origins of Life,* ed. A. Brack. Cambridge University Press: (New York) 219–36.

Ertem, G., and J. P. Ferris. 1997. Template-directed synthesis using the heterogeneous templates produced by montmorillonite catalysis. A possible bridge between the prebiotic and RNA worlds. *J. Am. Chem. Soc.* 119: 7197–7201.

Ferris, J. P. 1987. Prebiotic synthesis: Problems and challenges. *Cold Spring Harbor Symposia on Quantitative Biology* 52:29–35.

———. 1992. Marine hydrothermal systems and the origin of life: Chemical markers of prebiotic chemistry in hydrothermal systems. *Origins Life Evol. Biosphere* 22:109–34.

Ferris, J. P.; A. R. Hill, Jr.; R. Liu; and L. E. Orgel. 1996. Synthesis of long prebiotic oligomers on mineral surfaces. *Nature* 381:59–61.

Hanczyc, M. M.; S. M. Fujikawa; and J. W. Szostak. 2003. Experimental models of primitive cellular compartments: Encapsulation, growth, and division. *Science* 302:618–22.

Hill, A. R. Jr.; C. Böhler; and L. E. Orgel. 1998. Polymerization on the rocks: Negatively-charged amino acids. *Origins Life Evol. Biosphere* 28:235–43; scheme 1 on p. 236.

Joyce, G. F., and L. E. Orgel. 1993. Prospects for understanding the origin of the RNA World. *The RNA World,* eds. R. F. Gesteland and J. F. Atkins. Cold Spring Harbor Laboratory Press: (Stony Brook, N.Y.) 1–25.

Katchalsky, A. 1973. Prebiotic synthesis of biopolymers on inorganic templates. *Naturwiss* 60:215–20.

Lee, D. H.; J. R. Granja; J. A. Martinez; K. Severin; and M. R. Ghadiri. 1996. A self-replicating peptide. *Nature* 382:525–28.

Liu, R., and L. E. Orgel. 1998. Polymerization on the rocks: amino acids and arginine. *Origins Life Evol. Biosphere* 28:245–57.

Maizels, N., and A. M. Weiner, eds. 1993. The genomic tag hypothesis: Modern viruses as molecular fossils of ancient strategies for genome replication. *The RNA World*. Cold Spring Harbor Laboratory Press: (Stony Brook, N.Y.). 577–602.

McCollum, T. M., G. Ritter, and B. R. T. Simoneit. 1999. Lipid synthesis under hydrothermal conditions. *Origins Life Evol. Biosphere* 29. In press.

Melnick, J. L. 1999. Taxonomy and classification of viruses. In *Manual of Clinical Microbiology*, 7th ed., eds. P. R. Murray, E. J. Bacon, M. A. Pfaller, F. C. Tenover, and R. H. Yolken. ASM Press: (Washington, D.C.). 835–42.

Miller, L. Stanley. 1955. Production of some organic compounds under possible primitive earth conditions. *J. Am. Chem. Soc.* 77(9):2351.

Nielsen, P. E. 1999. Peptide nucleic acid. A molecule with two identities. *Acc. Chem. Res.,* 32:624–30.

Prabahar, K. J., and J. P. Ferris. 1997. Adenine derivatives as phosphate-activating groups for the regioselective formation of 3',5'-linked oligoadenylates on montmorillonite: Possible phosphate-activating groups for the prebiotic synthesis of RNA. *J. Am. Chem. Soc.* 119:4330–37.

Reimann, R., and G. Zubay. 1999. Nucleoside phosphorylation: A feasible step in the prebiotic pathway to RNA. *Origins Life Evol. Biosphere* 29. In press.

Shapiro, R. 1987. *Origins: A Skeptic's Guide to the Creation of Life on Earth.* Bantam Books: (New York).

———. 1988. Prebiotic ribose synthesis. A critical analysis. *Origins Life Evol. Biosphere.* 18:71–85.

Steinbeck, C., and C. Richert. 1998. The role of ionic backbones in RNA structure: An unusually stable non-Watson-Crick duplex of a nonionic analog in an apolar medium. *J. Am. Chem. Soc.* 120:11576–580.

Szostak, J. W., and A. D. Ellington. 1993. In vitro selection of functional RNA sequences. *The RNA World*, eds. R. F. Gesteland and J. F. Atkins. Cold Spring Harbor Laboratory Press: (Stony Brook, N.Y.) 511–33.

Szostak, J.; D. Bartel; and P. Luisi. 2001. Synthesizing life. *Nature* 409:387–90.

Tawfik, D. S., and A. D. Griffiths. 1998. Man-made cell-like compartments for molecular evolution. *Nat. Biotechnol.* 16:652–56.

Tidona, C. A.; G. Darai; and C. Büchen-Osmond, eds. 2001. *The Springer Index of Viruses.* Springer-Verlag KG: (New York).

van Regenmortel, M. H. V.; van C. M. Fauquet; D. H. L. Bishop; E. Carstens; M. K. Estes; S. Lemon; J. Maniloff; M. A. Mayo; D. J. McGeoch; C. R. Pringle; and R. Wickner, eds. 2000. *Virus Taxonomy. Classification and Nomenclature of Viruses.* Seventh Report of the International Committee on Taxonomy of Viruses. Academic Press: (New York).

Zimmer, C. 2004. What came before DNA? *Discover* 34–41.

Zubay, G. 1994. A feasible prebiotic pathway to the purines. *Chemtracts-Biochem. Mol. Biol.* 5: 179–89.

———. 1998. Studies on the lead-catalyzed synthesis of aldopentoses. *Origins Life Evol. Biosphere* 28:13–26.

Zubay, G., and T. Mui. 2001. Prebiotic synthesis of nucleotides. *Origins Life Evol. Biosphere* 31:87–102.

Chapter 6. Are There Aliens Already on Earth?

Kelley, D. S., et al. 2001. An off-axis hydrothermal vent field near the Mid-Atlantic Ridge at 30°N. *Nature* 412:145.

Venter, C., et al. 2004. Environmental genome shotgun sequencing of the Sargasso Sea. *Science* 304:66–74.

Chapter 7. Panspermia: Why There May Be Aliens Throughout the Solar System

Arrhenius, S. 1908. *Worlds in the Making.* Harper and Brothers.

Cano, R. J., and M. K. Borucki. 1995. Revival and identification of bacterial spores in 25- to 40-million-year-old Dominican amber. *Science* 268:1060–64.

Crick, F. H., and L. E. Orgel. 1973. Directed panspermia. *Icarus* 19:341–46.

Dones, L.; B. Gladman; H. J. Melosh; W. B. Tonks; H. F. Levison; and M. Duncan. 1999. Dynamical lifetimes and final fates of small bodies: Orbit integrations vs Öpik calculations. *Icarus* 142:509–24.

Farinella, P.; C. Froeschlé; R. Gonczi; G. Hahn; A. Morbidelli; and G. B. Valsecchi. 1994. Asteroids falling into the sun. *Nature* 371:314–17.

Gladman, B. 1997. Destination: Earth: Martian meteorite delivery. *Icarus* 130:228–46.

Gladman, B., and Burns, J. A. 1996. Martian meteorite transfer: Simulation. *Science* 274:161–62.

Gladman, B. J.; J. A. Burns; M. Duncan; P. Lee; and H. F. Levison. 1996. The exchange of impact ejecta between the terrestrial planets. *Science* 271:1387–92.

Hazen, R. M., and E. Roedder. 2001. How old are bacteria from the Permian age? *Nature* 411:155.

Head, J. N.; H. J. Melosh; and B. A. Ivanov. 2002. Martian meteorite launch: High-speed ejecta from small craters. *Science* 298:1752–56.

Horneck, G. 1993. Responses of *Bacillus subtilis* spores to space environment: Results from experiments in space. *Orig. Life Evol. Biosph.* 23:37–52.

Kirschvink, J. L. 2002. Mars, panspermia, and the origin of life: Where did it all begin? [abstract]. In *Impacts and the Origin, Evolution, and Extinction of Life*, UCLA, Los Angeles: 25. Available at: http://www.ess.ucla.edu/rubey/index.html.

Levison, H. F., and M. J. Duncan. 1997. From the Kuiper belt to Jupiter-family comets: The spatial distribution of ecliptic comets. *Icarus* 127:13–32.

Mastrapa, Glanzberg, Head, Melosh and Nicholson. 2001. Survival of bacteria exposed to extreme acceleration: Implications for panspermia. *Earth and Planetary Science Letters* 189:1–8.

McSween, H. Y. 1985. SNC meteorites: Clues to Martian petrologic evolution? *Rev. Geophys.* 23:391–416.

Melosh, H. J. 1985. Ejection of rock fragments from planetary bodies. *Geology* 13:144–48.

———. 1988. The rocky road to panspermia. *Nature* 332:687–88.

———. 1994. Swapping rocks. *Planet. Rep.* 14:16–19.

Melosh, H. J., and W. B. Tonks. 1994. Swapping rocks: Ejection and exchange of surface material among the terrestrial planets. *Meteoritics* 28:398.

Mileikowsky, C.; F. Cucinotta; J. W. Wilson; B. Gladman; G. Horneck; L. Lindgren; H. J. Melosh; H. Rickman; M. J. Valtonen; and J. Q. Zheng. 2000. Natural transfer of viable microbes in space. Part 1: From Mars to Earth and Earth to Mars. *Icarus* 145:391–427.

Miller, S. L., and A. Lazcano. 1996. From the primitive soup to Cyanobacteria: It may have taken less than 10 million years. In *Circumstellar Habitable Zones*, ed. L. R. Doyle. Travis House Publications: (Menlo Park, Ca.) 93–404.

Mojzsis, S. J.; G. Arrhenius; K. D. McKeegan; T. M. Harrison; A. P. Nuttman; and C. R. L. Friend. 1996. Evidence for life on earth before 3,800 million years ago. *Nature* 384:55–59.

Öpik, E. J. 1951. Collision probabilities with the planets and the distribution of interplanetary matter. *Proc. R. Ir. Acad.* 54:165–99.

Schopf, J. W. 1993. Microfossils of the Early Archean Apex Chert: New evidence of the antiquity of life. *Science* 260:640–46.

Seaward, M. R. D.; T. Cross; and B. A. Unsworth. 1976. Viable bacterial spores recovered from an archeological excavation. *Nature* 261:407–08.

Vreeland, R. N.; W. D. Rosenzweig; and D. W. Powers. 2000. Isolation of a 250-million-year-old halotolerantbacterium from a primary salt crystal. *Nature* 407:897–900.

Weissman, P. R. 1982. Terrestrial impact rates for long- and short-period comets. In *Geological Implications of Impacts of Large Asteroids and Comets on the Earth*, Special Paper 190, eds. L. T. Silver and P. H. Schultz. Geological Society of America: (Washington, D.C.) 15–24.

Zheng, J. Q., and M. J. Valtonen. 1999. On the probability that a comet that has escaped from the solar system will collide with the earth. *Monthly Notices R. Astron. Soc.* 304, 579–82.

Chapter 8. Mercury and Venus

Cockell, C. S. 1999. Life on Venus. *Planetary Space Sci.* 47:1487–1501.

Colin, J., and J. F. Kasting. 1992. Venus: A search for clues to early biological possibilities. In *Exobiology in the Solar System.* NASA Technical Pub. 512: (Washington, D.C.).

Conrad, P. G., and K. H. Nealson. 2001. A non-Earthcentric approach to life detection. *Astrobiology* 1:15–24.

Kasting, J. F. 1988. Runaway and moist greenhouse atmospheres and the evolution of earth and Venus. *Icarus* 74:472–94.

Kasting, J. F.; D. P. Whitmire; and R. T. Reynolds. 1993. Habitable zones around main sequence stars. *Icarus* 101:108–28.

Sagan, C., and H. Morowitz. 1967. Life in the clouds of Venus. *Nature* 215: 1259–60.

Schulze-Makuch, D.; L. N. Irwin; and T. Irwin. 2002. Astrobiological relevance and feasibility of a sample collection mission to the atmosphere of Venus. *ESA Sp.*, 518:247–52.

Schulze-Makuch, D., and L. N. Irwin. 2004. *Life in the Universe: Expectations and Constraints.* Springer-Verlag: (New York).

Schulze-Makuch, D.; D. H. Grinspoon; O. Abbibas; L. N. Irwin; and M. A. Bullock. 2004. A sulfur-based survival strategy for putative phototropic life in the Venusian atmosphere. *Astrobiology* 4:1–8.

Seckbach, J., and W. F. Libby. 1970. Vegetative life on Venus: Investigations with algae which grow under pure CO_2 in hot acid media at elevated pressures. *Space Life Sciences* 2:121–43.

Chapter 9. Fossils on the Moon

Alvarez, L.; W. Alvarez; F. Asaro; and H. Michel. 1980. Extra-terrestrial cause for the Cretaceous-Tertiary extinction: *Science* 208:1094–1108.

Amari, S.; E. Anders; A. Virag; and E. Ziner. 1990. Interstellar graphite in meteorites. *Nature* 345:238–40.

Benítez, N., et al. 2002. Evidence for Nearby supernova explosions, *Physical Review Letters,* 88 081101-1-4 at 081101-3.

Covey, C.; S. Thompson; P. Weissman; and M. Maccracken. 1994. Global climatic effects of atmospheric dust from and asteroid or comet impact on earth. *Global and Planetary Change* 9:263–73.

Crutzen, P., and C. Brühl. 1996. Mass extinctions and supernova explosions, *Proc. Natl. Acad. Sci. USA* 93:1582–84.

Ellis, J., and D. Schramm. 1995. Could a supernova explosion have caused a mass extinction? *Proc. Natl. Acad. Sci.* 92:235–38.

Erwin, D. 1994. The Permo-Triassic extinction. *Nature* 367:231–36.

Hallam, A., and P. Wignall. 1997. *Mass Extinctions and Their Aftermath.* Oxford University Press: (New York).

Hsu, K., and J. Mckenzie. 1990. Carbon isotope anomalies at era boundaries: Global catastrophes and their ultimate cause. *Geol. Soc. Am. Special Paper* 247:61–70.

Knoll, A.; R. Bambach; D. Canfield; and J. Grotzinger. 1996. Comparative earth history and Late Permian mass extinction. *Science* 273:452–57.

Marshall, C., and P. Ward. 1996. Sudden and gradual molluscan extinctions in the latest Cretaceous of Western European Tethys. *Science* 274:1360–63.

Melosh, H. J. 1994. Swapping rocks: Exchange of surface material among the planets. *Planetary Report* 14:16–19.

Pope, K.; A. Baines; A. Ocampo; and B. Ivanov. 1994. Impact winter and the Cretaceous Tertiary extinctions: Results of a Chicxulub asteroid impact model. *Earth and Planetary Science Express* 128:719–25.

Rampino, M., and K. Caldeira. 1993. Major episodes of geologic change: Correlations, time structure and possible causes: *Earth Planet. Sci. Lett.* 114:215–27.

Raup, D. 1990. Impact as a general cause of extinction: A feasibility test. In *Global Catastrophes in Earth History*, eds. V. Sharpton and P. Ward, *Geol. Soc. Am. Special Paper* 247:27–32.

———. 1991. A kill curve for Phanerozoic marine species. *Paleobiology* 17:37–48.

Raup, D., and J. Sepkoski. 1984. Periodicity of extinction in the geologic past. *Proc. Nat. Acad. Sci.* A81:801–05.

Retallack, G. 1995. Permian-Triassic crisis on land. *Science* 267:77–80.

Ruderman, M. 1974. Possible consequences of nearby supernova explosions for atmospheric ozone and terrestrial life. *Science* 184:1079–81.

Schindewolf, O. 1963. Neokatastrophismus? Zeit. *Der Deutschen Geol. Gesell.* 114:430–45.

Schultz, P. 1994. Impact angle effects on global lethality. *Lunar and Planetary Institute Contribution* 825:108–10.

Sigurdsson, H.; S. D'hondt; and S. Carey. 1992. The impact of the Cretaceous-Tertiary bolide on evaporite terrain and generation of majorsulfuric acid aerosol. *Earth Planetary Science Letters* 109:543–59.

Stanley, S. 1987. *Extinctions.* W. H. Freeman: (New York).

Stanley, S., and Yang, X. 1994. A double mass extinction at the end of the Paleozoic era. *Science* 266:1340–44.

Ward, P. 1990. The Cretaceous/Tertiary extinctions in the marine realm: A 1990 perspective: *Geol. Soc. Am. Special Paper* 247:425–432.

———. 1994. *The End of Evolution:* Bantam Doubleday: (New York).

Ward, P., and W. Kennedy. 1993. Maastrichtian ammonites from the Biscay region (France and Spain). *Journal of Paleontology, Memoir* 34 67:58.

Chapter 10. Mars

Brakenridge, G. R., H. E. Newsom; and V. R. Baker. 1985. Ancient hot springs on Mars: Origins and paleoenvironmental significance of small Martian valleys. *Geology* 13:859–62.

Carl, M. H. 1996. *Water on Mars.* Oxford University Press: (New York).

Forget, F., and G. D. Pierrehumbert. 1997. Warming early Mars with carbon dioxide that scatters infrared radiation. *Science* 278:1273–76.

Griffith, L. L., and E. L. Shock. 1995. A geochemical model for the formation of hydrothermal carbonate on Mars. *Nature* 377:406–08.

Jakosky, B. M., and E. L. Shock. 1998. The biological potential of Mars, the early Earth, and Europa. *J. Geophys. Res.* 103:19359–64.

Kasting, J. F. 1984. Effects of high CO_2 levels on surface temperature and atmospheric oxidation state of the early earth. *Journal of Geophysical Research* 86:1147–58.

———. 1997. Warming early Earth and Mars. *Science* 276:1213–15.

———. 1993. New spin on ancient climate. *Nature* 364:759–60.

Kasting, J. F., and T. P. Ackerman. 1986. Climatic consequences of very high carbon dioxide levels in the earth's early atmosphere. *Science* 234:1383–85.

McKay, D. S.; E. K. Gibson Jr.; K. L. Thomas-Kepra; C. S. Romanek; S. J. Clemett; X. D. F. Chillier; C. R. Maechling; and R. N. Zare. 1996. Search for past life on Mars: Possible relic biogenic activity in Martian meteorite ALH 84001. *Science* 273:924–30.

McSween, H. Y., Jr. 1994. What we have learned about Mars from SNC meteorites. *Meteoritics* 29:757–79.

Price, P.B., and T. Sowers. 2004. Temperature dependence of metabolic rates for microbial growth, maintenance, and survival. *Proc. Natl. Acad. Sci.* 30 101(13): 4631–36.

Sagan, C., and C. Chyba. 1997. The early faint sun paradox: Organic shielding of ultraviolet-labile greenhouse gases. *Science* 276:1217–21.

Romanek, C. S.; M. M. Grady; I. P. Wright; D. W. Mittlefehldt; R. A. Socki; C. T. Pillinger; and E. K. Gibson, Jr. 1994. Record of fluid rock interactions on Mars from the meteorite ALH 84001. *Nature* 372:655–57.

Shock, E. L., and M. D. Schulte. 1997. Hydrothermal systems as locations of organic synthesis on the early Earth and Mars. *Orig. Life Evol. Biosphere.*

Stetter, K. O. 1996. *Hyperthermophiles in Ecosystems on Earth (and Mars?)*, eds. G. R. Bock and J. A. Goode. John Wiley and Sons: (New York) 1–18.

Treiman, A. H. 1995. A petrographic history of Martian meteorite ALH 84001: Two shocks and an ancient age. *Meteoritics* 30:294–302.

Watson, L. L.; I. D. Hutcheon; S. Epstein; and E. M. Stolper. 1994. Water on Mars: Clues from deuterium/hydrogen and water contents of hydrous phases in SNC meteorites. *Science* 265:86–90.

Chapter 11. Europa

Chyba, C. F. 2000. Energy for microbial life on Europa: A radiation-driven ecosystem on Jupiter's moon is not beyond the bounds of possibility. *Nature* 403:381–82.

Chyba, C. F., and C. Phillips. 2001. Possible ecosystems and the search for life on Europa. *PNAS* 98:801–04.

Gaidos, E. J.; K. H. Nealson; and J. L. Kirschvink. 1999. Life in ice-covered oceans. *Science* 284:1631–33.

Gilichinsky, D. A. 2002. Permafrost model of extraterrestrial habitat. In *Astrobiology: The Quest for the Conditions of Life,* eds. G. Horneck and C. Baumstark-Khan. Springer: (New York) 125–42.

Gilichinsky, D. A., V. S. Soina, and M. A. Petrova. 1993. Cryoprotective properties of water in the earth cryolithosphere and its role in exobiology. *Orig. Life Evol. Biosph.* 23:65–75.

Greenberg, R. 2002. Tides and the biosphere of Europa. *Am. Sci.* 90:48–55.

Greenberg, R., and P. Geissler. 2002. Europa's dynamic icy crust. *Meteor. Planet. Sci.* 37:1685–1710.

Greenberg, R.; B. R. Tufts; P. Geissler; and G. V. Hoppa. 2002. Europa's crust and ocean: How tides create a potentially habitable physical setting. In *Astrobiology: The Quest for the Conditions of Life,* eds. G. Horneck and C. Baumstark-Khan. Springer: (New York) 111–24.

Hawes, I.; R. Smith; C. Howard-Williams; and A. M. Schwarz. 1999. Environmental conditions during freezing, and response of microbial mats in ponds of the McMurdo Ice Shelf, Antarctica. *Antarctic Sci.* 11:198–208.

Helmke, E., and H. Weyland. 1995. Bacteria in sea ice and underlying water of the eastern Weddell Sea in midwinter. *Mar. Ecol. Progr. Ser.* 117:269–87.

Irwin, L., and D. Schulze-Makuch. 2004. Europa: Strategy for modeling putative multilevel ecosystem on Europa. *Astrobiology* 3:813–21.

Jouzel, J.; J. R. Petit; R. Souchez; N. I. Barkov; V. Y. Lipenkov; D. Raynaud; M. Stievenard; N. I. Vassiliev; V. Verbeke; and F. Vimeux, 1999. More than 200 meters of lake ice above subglacial Lake Vostok, Antarctica. *Science* 286:2138–41.

Junge, K.; C. Krembs; J. W. Deming; A. Stierle; and H. Eicken. 2001. A microscopic approach to investigate bacteria under in-situ conditions in sea-ice samples. *Ann. Glaciol.* 33:304–10.

Junge, K.; H. Eicken; and J. W. Deming. 2004. Bacterial activity at 22 to 220°C in Arctic wintertime sea ice. *Appl. Environ. Microbiol.* (in press).

Kapitsa, A. P.; J. K. Ridley; G. de Q. Robin; M. J. Siegert; and I. A. Zotikov. 1996. A large deep freshwater lake beneath the ice of central East Antarctica. *Nature* 381:684–86.

Kargel, J. S. 1991. Brine volcanism and the interior structures of asteroids and icy satellites. *Icarus* 94:368–90.

———. 2001. Roles of Europa's stratified crust and ocean in diapirism and melt-through. In *Europa Focus Group Workshop* 2, ed. R. Greeley. Arizona State University: (Tempe, Az.) 19–20.

Kargel, J. S.; J. Kaye; J. W. Head III; G. M. Marion; R. Sassen; J. Crowley; O. Prieto; S. A. Grant; and D. Hogenboom. 2000. Europa's salty crust and ocean: Origin, composition, and the prospects for life. *Icarus* 148:226–65.

Kargel, J. S.; J. W. Head III; D. L. Hogenboom; K. K. Khurana; and G. M. Marion. 2001. The system sulfuric acidmagnesium sulfate-water: Europa's ocean properties related to thermal state [abstract 2138]. In *Lunar and Planetary Science Conference XXXII.* Lunar and Planetary Institute.

Karl, D. M.; D. F. Bird; K. Björkman; T. Houlihan; R. Shackelford; and L. Tupas. 1999. Microorganisms in the accreted ice of Lake Vostok, Antarctica. *Science* 286:2144–47.

Kato, C.; L. Li; Y. Nogi; Y. Nakamura; J. Tamaoka; and K. Horikoshi. 1998. Extremely barophilic bacteria isolated from the Mariana Trench, Challenger Deep, at a depth of 11,000 meters. *Appl. Environ. Microbiol.* 64:1510–13.

Kaye, J. Z., and J. A. Baross. 2000. High incidence of halotolerant bacteria in Pacific hydrothermal-vent and pelagic environments. *FEMS Microbiol. Ecol.* 32:249–60.

Kelly, D. S. 2001. Black smokers: Incubators on the seafloor. In *Earth: Inside and Out,* ed. E. A. Mathez, New Press: (New York) 183–89.

Kempe, S., and E. T. Degens. 1985. An early soda ocean? *Chem. Geol.* 53:95–108.

Kempe, S., and J. Kazmierczak. 2002. Biogenesis and early life on earth and Europa: Favored by an alkaline ocean? *Astrobiology* 2:123–30.

Kennedy, A. D. 1993. Water as a limiting factor in the Antarctic terrestrial environment: A biogeographical synthesis. *Arctic Alpine Res.* 25:308–15.

Lunine, J., and R. Lorenz. 1997. Light and heat in cracks on Europa: Implications for prebiotic synthesis. *Lunar. Plan. Sci.* 28:855–56.

Madigan, M. T., and A. Oren. 1999. Thermophilic and halophilic extremophiles. *Curr. Opin. Microbiol.* 2:265–69.

Marion, G. M. 1997. A theoretical evaluation of mineral stability in Don Juan Pond, Wright Valley, Victoria Land. *Antarctic Sci.* 9:92–99.

———. 2002. A molal-based model for strong acid chemistry at low temperatures (200 to 298 K). *Geochim. Cosmochim. Acta* 66:2499–2516.

Mazur, P. 1980. Limits to life at low temperatures and at reduced water contents and water activities. *Orig. Life* 10:137–59.

McCollom, T. M. 1999. Methanogenesis as a potential source of chemical energy for primary biomass production by autotrophic organisms in hydrothermal systems on Europa. *J. Geophys. Res.* 104:30729–42.

McCord, T. B.; G. B. Hansen; D. L. Matson; T. V. Johnson; J. K. Crowley; F. P. Fanale; R. W. Carlson; W. D. Smythe; P. D. Martin; C. A. Hibbitts; J. C. Granahan; and A. Ocampo. 1999. Hydrated salt minerals on Europa's surface from the Galileo Near-Infrared Mapping Spectrometer (NIMS) investigation. *J. Geophys. Res.* 104:11827–51.

McKinnon, W. B. 1997. Sighting the seas of Europa. *Nature* 386:765–67.

Meyer, G. H.; M. B. Morrow; O. Wyss; T. E. Berg; and J. L. Littlepage. 1962. Antarctica: The microbiology of an unfrozen saline pond. *Science* 138:1103–04.

Millero, F. J., and M. L. Sohn. 1992. *Chemical Oceanography,* CRC Press: (Boca Raton, Fla.).

O'Brien, D. P.; P. Geissler; and R. Greenberg. 2002. A melt-through model for chaos formation on Europa. *Icarus* 156:152–61.

Pappalardo, R. T., et al. 1999. Does Europa have a subsurface ocean? Evaluation of the geological evidence. *J. Geophys. Res.* 104:24015–55.

Pedersen, K. 1993. The deep subterranean biosphere. *Earth-Sci. Rev.* 34:243–60.

Phillips, C. B., and C. F. Chyba. 2001. Impact and sputtering rates on Europa: Competing processes in the production, destruction, and preservation of biogenic compounds. In *Europa Focus Group Workshop 1,* ed. R. Greeley, NASA-Ames: 22.

Phoenix, V. R.; K. O. Konhauser; D. G. Adams; and S. H. Bottrell. 2001. Role of biomineralization as an ultraviolet shield: Implications for Archean life. *Geology* 29:823–26.

Pierazzo, E., and C. F. Chyba. 2002. Cometary delivery of biogenic elements to Europa. *Icarus* 157:120–27.

Pledger, R. J., and J. A. Baross. 1991. Preliminary description and nutritional characterization of a chemoorganotrophic archaeobacterium growing at temperatures of up to 110°C isolated from a submarine hydrothermal vent environment. *J. Gen. Microbiol.* 137:203–11.

Price, P. B. 2000. A habitat for psychrophiles in deep Antarctic ice. *Proc. Natl. Acad. Sci. USA* 97:1247–51.

Priscu, J. C., et al. 1999. Geomicrobiology of subglacial ice above Lake Vostok, Antarctica. *Science* 286:2141–44.

Psenner, R., and B. Sattler. 1998. Life at the freezing point. *Science* 280:2073–74.

Reynolds, R. T.; S. W. Squyres; D. S. Colburn; and C. P. McKay. 1983. On the habitability of Europa. *Icarus* 56:246–54.

Rivkina, E. M.; E. I. Friedmann; C. P. McKay; and D. A. Gilichinsky. 2000. Metabolic activity of permafrost bacteria below the freezing point. *Appl. Environ. Microbiol.* 66:3230–33.

Schroeter, B., and C. Scheidegger. 1995. Water relations in lichens at subzero temperatures: Structural changes and carbon dioxide exchange in the lichen *Umbilicaria aprina* from continental Antarctica. *New Phytol.* 131: 273–85.

Schulze-Makuch, D., and L. N. Irwin. 2002. Energy cycling and hypothetical organisms in Europa's ocean. *Astrobiology* 2:105–21.

Segerer, A. H.; S. Burggraf; G. Fiala; G. Huber; R. Huber; U. Pley; and K. O. Stetter. 1993. Life in hot springs and hydrothermal vents. *Orig. Life Evol. Biosph.* 23:77–90.

Sharma, A.; J. H. Scott; G. D. Cody; M. L. Fogel; R. M. Hazen; R. J. Hemley; and W. T. Huntress. 2002. Microbial activity at gigapascal pressures. *Science* 295:1514–16.

Soare, R.; W. Pollard; and D. Green. 2001. Deductive model proposed for evaluating terrestrial analogues. *EOS* 82, 501.

Soina, V. S.; E. A. Vorobiova; D. G. Zvyagintsev; and D. A. Gilichinsky. 1995. Preservation of cell structures in permafrost: A model for exobiology. *Adv. Space Res.* 15:237–42.

Spencer, J. R.; L. K. Tamppari; T. Z. Martin; and L. D. Travis. 1999. Temperatures on Europa from Galileo photopolarimeter-radiometer: Nighttime thermal anomalies. *Science* 284:1514–16.

Stapleton, R. D.; D. C. Savage; G. S. Sayler; and G. Stacey. 1998. Biogradation of aromatic hydrocarbons in an extremely acidic environment. *Appl. Environ. Microbiol.* 64:4180–84.

Stevenson, D. 2000. Europa's ocean: The case strengthens. *Science* 289:1305–07.

Stone, R. 1999. Permafrost comes alive for Siberian researchers. *Science* 286:36–37.

Thomson, R. E., and J. R. Delaney. 2001. Evidence for a weakly stratified Europan oceans sustained by seafloor heat flux. *J. Geophys. Res.* 106:12355–65.

Ventosa, A.; J. J. Nieto; and A. Oren. 1998. Biology of moderately halophilic aerobic bacteria. *Microbiol. Mol. Biol. Rev.* 62:504–44.

Vogel, G. 1999. Expanding the habitable zone. *Science* 286:7071.

Zahnle, K.; L. Dones; and H. F. Levison. 1998. Cratering rates on the Galilean satellites. *Icarus* 136:202–22.

Zolotov, M. Y., and E. L. Shock. 2001. Composition and stability of salts on the surface of Europa and their oceanic origin. *J. Geophys. Res.* 106:32815–27.

Chapter 12. Titan

Chamberlain, J. W., and D. M. Hunten. 1987. *Theory of Planetary Atmospheres.* Academic Press: (New York).

Fortes, A. D. 2000. Exobiological implications of a possible ammonia-water ocean inside Titan. *Icarus* 146:444–52.

Gautier, D., and B. J. Conrath. 1995. The troposphere of Neptune. In *Neptune and Triton,* ed. D. P. Cruikshank. University of Arizona Press: (Tempe, Az.) 547–611.

Gonzalez, G.; D. Brownlee; and P. Ward. 2001. The galactic habitable zone: Galactic chemical evolution. *Icarus* 152:185–200.

Grasset, O.; C. Sotin; and F. Deschamps, F. 2000. On the internal structure and dynamics of Titan. Planet. *Spac Sci.* 48:617–36.

Guillot, T. 1999. Interiors of giant planets inside and outside the solar system. *Science* 286:72–76.

Hart, M. H. 1979. Habitable zones about main sequence stars. *Icarus* 37:351–57.

Kirk, R. L.; L. A. Soderblom; R. H. Brown; S. W. Kieffer; and J. S. Kargel. 1995. Triton's plumes: Discovery, characteristics and models. In *Neptune and Triton,* ed. D. Cruickshank. University of Arizona Press: (Tempe, Az.) 949–89.

Lara, L. M.; R. D. Lorenz; and R. Rodrigo. 1994. Liquids and solids on the surface of Titan: Results of a new photochemical model. *Planet. Space Sci.* 42:5–14.

Lorenz, R. D. 1994. Crater lakes on Titan: Rings, horseshoes, bullseyes. *Planet. Space Sci.* 42:1–4.

————. 2002. Thermodynamics of geysers: Application to Titan. *Icarus* 156: 176–83.

Lorenz, R. D., and J. I. Lunine. 2002. Titan's snowline. *Icarus* 158:557–59.

Lorenz, R., and J. Mitton. 2002. *Lifting Titan's Veil.* Cambridge University Press: (New York).

Meier, R.; B. A. Smith; T. C. Owen; and R. J. Tervile. 2000. The surface of Titan from NICMOS observations with the Hubble space telescope. *Icarus* 145:462–73.

Pietrogrande, M. C.; P. Coll; R. Sternberg; C. Szopa; R. Navarro-Gonzalez; C. Vidal-Madjar; and F. Dondi. 2001. Analysis of complex mixtures recovered from space missions: Statistical approach to the study of Titan atmosphere analogues (tholins). *Journal of Chromatography* 939:69–77.

Raulin, F.; P. Bruston; and P. Pailous. 1995. The low temperature organic chemistry of Titan's geofluid. *Adv. Space Res.* 15:321–33.

Rothery, D. A. 1992. *Satellites of the Outer Planets: Worlds in Their Own Right.* Clarendon Press: (New York).

Sagan, C.; N. B. Khare; L. E. Bandurski; and N. Batholomew. 1978. Ultraviolet-photoproduced organic solids synthesized under simulated Jovian conditions: Molecular analysis. *Science* 199:1199–1201.

Sagan, C., and Khare N. B. 1979. Tholins: organic chemistry of interstellar grains and gas. *Nature* 277:102–07.

Sagan, C.; W. R. Thompson; and B. N. Khare. 1992. Titan: A laboratory for prebiological organic chemistry. *Accounts of Chemical Research* 25:286–92.

Shenk, P. M. 2002. Thickness constraints on the icy shells of the Galilean satellites from a comparison of crater shapes. *Nature* 417:419–21.

Showman, A. P., and R. Manotra. 1999. The Galilean satellites. *Science* 286:77–84.

Sleep, N. H., and K. Zahnle. 2001. Carbon dioxide cycling and implications for climate on ancient earth. *J. Geophys. Res. Planets* 106:1373–99.

Soderblom, L. A.; S. W. Kieffer; T. L. Becker; R. H. Brown; A. F. Cook II; C. J. Hansen; T. V. Johnson; R. L. Kirk; and E. M. Shoemaker. 1990. Subsurface liquid N2 suggested by geyser-like eruptions on Triton. *Science* 250:410–15.

Taylor, F. W., and A. Coustenis. 1998. Titan in the solar system. *Planet. Space Sci.* 46:1085–97.

Williams, D. M.; J. F. Kasting; and R. A. Wade. 1997. Habitable moons around extrasolar giant planets. *Nature* 385:234–36.

Chapter 13. Implications, Ethics, and Dangers

DeVincenzi, D. 1989. Proposed planetary protection guidelines for sample return missions 19 (3-5) origins of life and evolution of the biosphere (abstracts of ISSOL meeting, Prague, 1989) 485.

DeVincenzi, D., and J. R. Bagby. 1981. Orbiting quarantine facility, National Aeronautics and Space Administration Doc. No. SP-454.

DeVincenzi, D. H.; P. Klein; and J. R. Bagby. 1991. Protection issue and future Mars mission, National Aeronautics and Space Administration Doc. No. CP-10086.

DeVincenzi, D. H.; P. Klein, and J. R. Bagby, eds. 1994. Planetary Protection Issues and Future Mars Missions, National Aeronautics and Space Administration Supplement Doc. No. NAS 1.55:10086.

Horowitz, N. R.; P. Sharp; and R. W. Davies. 1967. Planetary contamination I: The problem and the agreements, *Science* 55 1501.

National Research Council (U.S.). 1992. Task Group on Planetary Protection, Biological contamination of Mars, issues and recommendations (NASA CR-190819. NAS 1.26:190819).

Phillips, C. 1974. The planetary quarantine program—origins and achievements (1956–1973), (NASA SP-4902; Sup. Doc. No. NAS 1.21:4902).

Pilcher, C. 1992. From a lunar outpost to Mars: Science, policy and the U.S. space exploration initiative 12(4) *Adv. Space Res.* 91

Planetary protection implementation on future Mars lander missions: U.S./Russian Workshop. 1992. Palo Alto, Calif., National Aeronautics and Space Administration Sup. Doc. No. NAS 1.55:3216, eds. R. Howell, D. H. DeVincenzi.

Race, M. 1994. Societal Issues as Mars missions impediments: Planetary protection and contamination concerns 15 *Adv. Space Res.* 285 19.

———. 1995. Integration of planetary protection activities, NASA Contractor Report NASA CR-200223, Supplement Document No. NAS 1.26:200223.

———. 1996. Planetary protection, legal ambiguity and the decision making process for Mars sample return 18 *Adv. Space Res.* 345.

Reynolds, O. 1973. Developments in the analysis of planetary quarantine requirements, XI *Life Sci. & Space Res.* 3.

Rummel, J. 1992. Planetary protection policy 12 (4) *Adv. Space Res.* 129.

Sterns, P., and L. I. Tennen. 1989. Recent developments in the planetary protection policy: Is the outer space environment at risk? *Proc. 32nd Colloquium on the Law of Outer Space.* 163.

Index